Current Topics in Microbiology and Immunology

Volume 317

Scott K. Dessain

Editor

Human Antibody Therapeutics for Viral Disease

 Springer

Scott K. Dessain, MD, PhD
Cardeza Foundation for Hematologic Research
Kimmel Cancer Center
Thomas Jefferson University
Curtis Building 812
1015 Walnut St.
Philadelphia, PA 19107
USA
E-mail: scott.dessain@jefferson.edu

Cover Illustration: Decoration of the E16 neutralizing monoclonal antibody on the West Nile virion. The E16 structural epitope is mapped in magenta onto the cyro-electron microscopic reconstruction of the West Nile virion. Domains I, II, and III of West Nile virus E protein are represented in red, yellow, and blue, respectively.

ISBN 978-3-540-72144-4   e-ISBN 978-3-540-72146-8

Current Topics in Microbiology and Immunology ISSN 0070-217X

Library of Congress Catalog Number: 72-152360

© 2008 Springer-Verlag Berlin Heidelberg

Cover design: WMX Design GmbH, Heidelberg

Printed on acid-free paper

9 8 7 6 5 4 3 2 1

springer.com

Preface

Editor's Introduction

Although the utility of human antibodies as medical therapeutics for cancer and immune diseases has been well-established, it is only beginning to be realized for the treatment of viral infectious diseases. Polyclonal immunoglobulins have long been used for some viral diseases, but they have limited potency and disease scope. It should theoretically be possible to create monoclonal or oligoclonal antibody preparations that capture the essential curative functions of the humoral immune response to viral pathogens, yet only a single humanized monoclonal antibody (pavilizumab) has been approved as a viral countermeasure. Reliable technologies for creating human or humanized antibodies with defined viral antigen specificities are well-established. Accordingly, current antibody development efforts are focused on identifying and cloning the particular antibodies that contain the fundamental curative potency of the polyclonal humoral immune response.

The limited spectrum of available viral antibody therapeutics may be attributed, in part, to the unique technical challenges that each virus presents with regard to its life cycle, capacity for spontaneous mutation, and clinical manifestations. The articles in this volume address these issues from different perspectives. For instance, the clinical value of an antibody may depend on the genetic stability of the epitope it recognizes, the prevalence of that epitope in the viral population, the kinetics of the antibody/antigen interaction, the functional implications of epitope binding, and the interactions between the antibody and the immune system, which are often mediated by the Fc portion of the antibody. In some settings, a single antibody may be sufficient, whereas in others a combination of 2 or more may be preferred.

The articles in this volume have been selected to demonstrate the progress in the development of human antibody therapeutics for viral disease. Keck et al. review the nature of the immune response to the Hepatitis C virus (HCV) and the details of viral neutralization by antibodies, providing a conceptual model for the clinical use of HCV-specific antibodies. Huber et al. summarize the initial clinical experiences with antibody therapeutics for Human Immunodeficiency Virus (HIV), which can be targeted to either the HIV virion or to host cell proteins. A discussion of the breadth immune strategies that is required to control human rabies is provided by Nagarajan et al., with a particular focus on India and other developing countries in which rabies is endemic. The development of pavilizumab for RSV prophylaxis is reviewed in Wu et al., in addition to results of antibody optimization studies that provide surprising insights and have broad general implications for anti-viral antibody engineering. Melhop and Diamond explicate the biology of West Nile Virus (WNV) as a general model for flaviviruses, while using their cloned antibodies as a springboard to consider the mechanisms of WNV neutralization. The volume concludes with a description of methods to clone human antibodies in their native configurations, which access a class of antibodies that differ from those obtained by recombinant DNA or transgenic mouse methods.

The articles in this volume are definitive and comprehensive reviews written by thought leaders who have sought to define the principles of viral neutralization by human antibodies. They explore and anticipate the obstacles and opportunities that will be encountered as the power of human antibodies is harnessed to address the vast, un-met need for effective anti-viral therapeutics.

Philadelphia, PA   Scott Dessain
March 2007

Contents

Contributors

Adekar, S.P.
Cardeza Foundation for Hematologic Research, Kimmel Cancer Center, Thomas
Jefferson University, Curtis Building 812, 1015 Walnut St., Philadelphia,
PA 19107, USA, sharad.adekar@jefferson.edu

Ball, J.K.
Institute of Infection, Immunity and Inflammation, School of Molecular Medical
Sciences, The University of Nottingham, Queen's Medical Centre, Nottingham
NG7 2UH, UK

Berry, J.D.
Monoclonal Antibody and Bioforensic Development Sections, National
Microbiology Laboratory, Public Health Agency of Canada (PHAC), 1015 Arlington
St., Winnipeg, Manitoba R3E 3R2, Canada

Dessain, S.K.
Cardeza Foundation for Hematologic Research, Kimmel Cancer Center, Thomas
Jefferson University, Curtis Building 812, 1015 Walnut St., Philadelphia,
PA 19107, USA, scott.dessain@jefferson.edu

Diamond, M.S.
Departments of Medicine, Molecular Microbiology, Pathology & Immunology,
Washington University School of Medicine, 660 South Euclid Avenue, Box 8051,
St. Louis, MO 63110, USA, diamond@borcim.wustl.edu

Foung, S.K.H.
Department of Pathology, Stanford Medical School Blood Center, 800 Welch
Road, Palo Alto, CA, 94304, USA, sfoung@leland.stanford.edu

Huber, M.
Division of Infectious Diseases, University Hospital Zurich, Rämistrasse 100,
8091 Zurich, Switzerland

Keck, Z.Y.
Department of Pathology, Stanford Medical School Blood Center, Palo Alto, CA,
94304, USA

Kiener, P.A.
MedImmune, Inc., One MedImmune Way, Gaithersburg, MD 20878, USA

Lai, M.M.C.
Department of Molecular Microbiology and Immunology, University of Southern
California, Keck School of Medicine, 2011 Zonal Avenue, Los Angeles,
CA 90033, USA

Losonsky, G.A.
MedImmune, Inc., One MedImmune Way, Gaithersburg, MD 20878, USA

Machida, K.
Department of Molecular Microbiology and Immunology, University of Southern
California, Keck School of Medicine, 2011 Zonal Avenue, Los Angeles,
CA 90033, USA

Mehlhop, E.
Departments of Medicine, Molecular Microbiology, Pathology & Immunology,
Washington University School of Medicine, 660 South Euclid Avenue, Box 8051,
St. Louis, MO 63110, USA

Nagarajan, T.
Indian Immunologicals Limited Gachibowli Post, Hyderabad, India,
drtnraj@rediffmail.com

Olson, W.C.
Progenics Pharmaceuticals, Inc., 777 Old Saw Mill River Road, Tarrytown,
NY, 10591 USA

Patel, A.H.
MRC Virology Unit, Institute of Virology, University of Glasgow, Church Street,
Glasgow G11 5JR, UK

Pfarr, D.S.
MedImmune, Inc., One MedImmune Way, Gaithersburg, MD 20878, USA

Rangarajan, P.N.
Indian Institute of Science, Bangalore, India

Rupprecht, C.E.
Centers for Disease Control and Prevention, Atlanta, GA, USA

Srinivasan, V.A.
Indian Immunologicals Limited Gachibowli Post, Hyderabad, India,
srini@indimmune.com

Thiagarajan, D.
Indian Immunologicals Limited Gachibowli Post, Hyderabad, India

Trkola, A.
Division of Infectious Diseases, University Hospital Zurich, Rämistrasse 100,
8091 Zurich, Switzerland, alexandra.trkola@usz.ch

Wu, H.
MedImmune, Inc., One MedImmune Way, Gaithersburg, MD 20878, USA,
wuh@medimmune.com

Therapeutic Control of Hepatitis C Virus: The Role of Neutralizing Monoclonal Antibodies

Z.Y. Keck, K. Machida, M.M.C. Lai, J.K. Ball, A.H. Patel,
and S.K.H. Foung(✉)

Abstract Liver failure associated with hepatitis C virus (HCV) accounts for a substantial portion of liver transplantation. Although current therapy helps some patients with chronic HCV infection, adverse side effects and a high relapse rate are major problems. These problems are compounded in liver transplant recipients as reinfection occurs shortly after transplantation. One approach to control reinfection is the combined use of specific antivirals together with HCV-specific antibodies. Indeed, a number of human and mouse monoclonal antibodies to conformational and linear epitopes on HCV envelope proteins are potential candidates, since they have high virus neutralization potency and are directed to epitopes conserved across

S.K.H. Foung
Department of Pathology, Stanford Medical School Blood Center, 3373 Hillview Avenue, Palo Alto, CA, 94304, USA
e-mail: sfoung@leland.stanford.edu

S.K. Dessain (ed.) *Human Antibody Therapeutics for Viral Disease. Current Topics in Microbiology and Immunology 317.*
© Springer-Verlag Berlin Heidelberg 2008

diverse HCV genotypes. However, a greater understanding of the factors contributing to virus escape and the role of lipoproteins in masking virion surface domains involved in virus entry will be required to help define those protective determinants most likely to give broad protection. An approach to immune escape is potentially caused by viral infection of immune cells leading to the induction hypermutation of the immunoglobulin gene in B cells. These effects may contribute to HCV persistence and B cell lymphoproliferative diseases.

1 Introduction

Over 170 million people worldwide are infected with hepatitis C virus (HCV). A large proportion of these individuals will harbor a chronic infection that eventually leads to the development of severe liver disease, liver failure, and hepatocellular carcinoma (Major et al. 2001). In the United States, 40% of patients needing liver transplantation have HCV-associated diseases. While current therapy with pegylated interferon and ribavirin is effective in some patients with chronic HCV hepatitis, adverse side effects and a high relapse rate, particularly in individuals infected with genotype 1 virus, are major problems. These problems are compounded in liver transplant recipients, where HCV reinfection occurs shortly after transplantation and antiviral treatment can only start at a later time point (Biggins and Terrault 2005). Allograft failure due to HCV reinfection is the most common cause of retransplantation and death (Charlton et al. 1998). HCV reinfection occurs immediately after transplantation and viral replication starts within hours of surgery (Garcia-Retortillo et al. 2002). Risk factors associated with recurrent and severe liver disease include donor and recipient age, acute rejection, human cytomegalovirus infection, and higher levels of HCV viral load. Serum viral load increases rapidly and peaks by 4 months after transplantation with titers up to 20 times higher than pretransplant levels at 1 year after transplantation (Everhart et al. 1999). Treatment of acute rejection with steroids leads to further increases in viral load. Combined treatment with pegylated interferon and ribavirin is more effective than interferon alone (Biggins and Terrault 2005), but combined therapy is associated with more than 50% of patients needing to stop treatment because of ribavirin-associated anemia. The occurrence of HCV reinfection in liver transplant recipients indicates that the prior immunity in HCV patients does not provide protection. Clearly, more effective strategies are needed for treating and preventing HCV reinfection among liver transplant recipients, and more understanding on the mechanisms of HCV immune escape is needed.

One possibility to control posttransplantation HCV reinfection is the combined use of specific antivirals together with HCV-specific antibodies. This type of approach is widely used for the control of hepatitis B virus (HBV) after liver transplantation, where the standard of care is pretransplant treatment with lamivudine for at least 4 weeks followed by posttransplant treatment with combined lamivudine and HBV immunoglobulin (Schreibman and Schiff 2006). When lamivudine or other nucleoside analogs are used alone, emergence of HBV escape mutants is a major concern (Perrillo et al. 2001; Fung et al. 2006). Similarly for HCV, a number of small molecule inhibitors of HCV NS3-4A protease and NS5B polymerase are in advanced stage clinical studies, potentially providing new therapeutic options (reviewed in De Francesco and Migliaccio 2005). However, a growing concern is that a high and error-prone viral replication will eventually lead to resistance to these antivirals. Our proposal is that virus neutralizing (Vn) antibodies when used in combination with antivirals could achieve the needed therapeutic outcome and minimize escape virus mutants. Selected antibodies ideally should be broadly reactive to different HCV genotypes, each inhibiting at different steps of virus entry, and be synergistic in their ability to control virus infection. At least two antibodies to different epitopes are proposed to minimize the concern of escape mutants. We and others have shown that the majority of HCV antibodies with the broadest and most potent Vn activities recognize conformational epitopes (Habersetzer et al. 1998; Allander et al. 2000; Ishii et al. 1998). Nonetheless, Vn monoclonal antibodies (MAbs) to conserved linear epitopes have been identified and should be considered as therapeutic candidates (Owsianka et al. 2001, 2005; Tarr et al. 2006). While antibodies offer an as yet under-explored avenue for HCV treatment, there are a number of factors that will influence the efficacy of this approach and therefore dictate the most appropriate strategy. These factors are considered below.

2 Evolutionary Dynamics of HCV Envelope Genes

HCV can be classified into six genetically distinct genotypes and further subdivided into a large number of subtypes, of which the six major genotypes differ by approximately 30% and the subtypes differ by 20%–25% at the nucleotide level (Simmonds et al. 2005). A significant challenge for vaccine and immunotherapeutic development is the identification of protective epitopes conserved in the majority of viral genotypes and subtypes. This problem is compounded by the fact that the envelope E1E2 proteins, the natural targets for the Vn response, are two of the most variable proteins (Bukh et al. 1995). The error-prone nature of the RNA-dependent RNA polymerase together with the high HCV replicative rate in vivo (Neumann et al. 1998) results in the production of viral quasispecies (Martell et al. 1994; Bukh et al. 1995). Quasispecies can respond to and overcome a variety of selective pressures including host immunity (Cooreman and Schoondermark-Van de Ven 1996; Cooreman and Schoondermark-Van de Ven 1996); chronic infection arises, at least in part, through the outgrowth of immune escape mutants (Frasca et al. 1999; Farci et al. 2000; Majid et al. 1999; Ray et al. 1999; Sullivan et al. 1999). The envelope glycoprotein genes display some of the highest levels of genetic heterogeneity, with E2 exhibiting greater variability at the quasispecies level than E1. Mean pair-wise genetic distances, calculated for translated HCV protein sequences spanning the entire E1E2 coding region, highlight regions of high and low variability and are distributed throughout all of the E1E2 genes (Fig. 1).

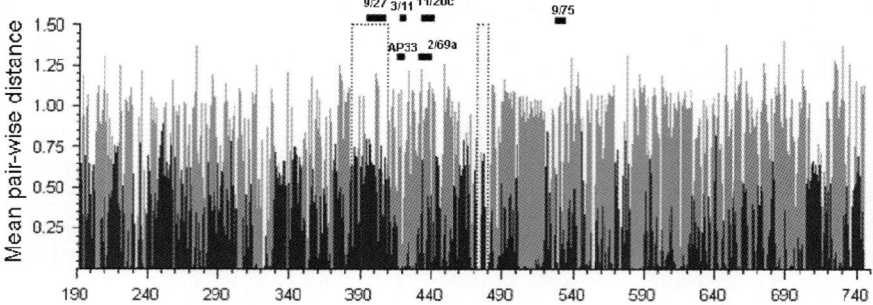

Fig. 1 Mean pair-wise distance plot for E1E2 amino acids and synonymous sites generated for HCV sequences representative of genotypes 1 through 6 (98 isolates, 543 codons). Mean p-distance at specific amino acid positions are represented by *black bars* in the foreground. Mean p-distance at synonymous sites in homologous codons are represented by *gray bars* in the background. Positions are relative to homologous codons in the H77 reference strain polyprotein. *Dotted columns* represent HVR1 and HVR2. The labeled *black bars* located above the diversity plot represent the relative positions of characterized Vn and non-Vn MAbs recognizing linear epitopes

    While high levels of amino acid diversity are evident, pair-wise distances calculated at synonymous (silent) sites revealed that a considerable proportion of genetic variation across HCV genotypes 1–6 is due to synonymous substitution. These high levels of synonymous site diversity observed throughout E1 and E2 indicate these genes are under strong purifying selection. Much of the current knowledge of adaptive evolution within the envelope genes during HCV chronic infection is based on averaged estimates of synonymous (d_S) and nonsynonymous (d_N) nucleotide substitution rates across very small regions of the envelope genes such as the first hypervariable region (HVR1; Curran et al. 2002; Gretch et al. 1996; Honda et al. 1994; McAllister et al. 1998; Smith 1999). However, these averaging types of analysis are too blunt to dissect out exact evolutionary mechanisms leading to antibody escape. Averaged d_N/d_S ratios across regions are highly conservative, as only a few codons within the protein may be under diversifying selection. The signal is therefore diluted in an overbearing purifying selective background that arises through strong functional constraint. To overcome analytical problems associated with differential selection across a region, the distribution of the d_N/d_S ratio (ω) can now be estimated for individual amino acids by assessing competing models of codon substitution within a maximum likelihood (ML) framework (Yang and Bielawski 2000). These ML methods have recently been applied to the identification of site-specific adaptive mutations in human immunodeficiency virus (HIV) *env* genes (Choisy et al. 2004) and partial E1E2 sequence data sets from individuals with chronic (Brown et al. 2005) and acute phases of HCV infection (Sheridan et al. 2004). The latter study extended earlier findings (Puig et al. 2004; Farci et al. 2000) revealing a statistically significant association between outcome of acute phase infection and the number of positively selected sites (Sheridan et al. 2004). Importantly, these methods can help pinpoint those residues and regions that the virus can easily change to help evade both cellular and humoral responses as well as those that are under strong functional constraint. This information is key to devising appropriate immunotherapeutic approaches, since utilizing antibodies that target highly conserved functionally constrained regions will minimize the likelihood of antibody escape and maximize breadth of action and therapeutic potential.

Analysis of viral evolution has shown that the amino terminus of the E2 envelope protein contains residues that have a very high propensity for adaptive change. Indeed, this is the only locus within E1E2 that harbors positively selected sites that are common across all genotypes (Fig. 2). This region, known as the first hypervariable region (HVR1), is the major determinant for strain-specific Vn antibody responses (Farci et al. 1994, 1996; Bartosch et al. 2003a; Rosa et al. 1996; Shimizu et al. 1994). There is increasing evidence that acute infection outcome is correlated to the rate and nature of nucleotide substitutions within the envelope genes (Farci et al. 2000; Ray et al. 1999; Sheridan et al. 2004). Patients harboring stable HVR1 quasispecies frequently resolve infection, while those with evidence of a rapidly evolving quasispecies develop chronic infection (Farci et al. 2000; Ray et al. 1999). Evolution of the viral quasispecies continues during the chronic phase (Brown et al. 2005; Gretch et al. 1996), and differences in evolutionary rates and disease severity in individuals with differing levels of immunocompetency highlight the importance of antibody responses in controlling the infection (Booth et al. 1998; Kumar et al. 1994). While there is strong evidence that antibodies directed to this region correlate with a beneficial outcome, the limited cross-reactivity of most HVR1-specific responses presents a serious problem to their wider therapeutic potential. A recent

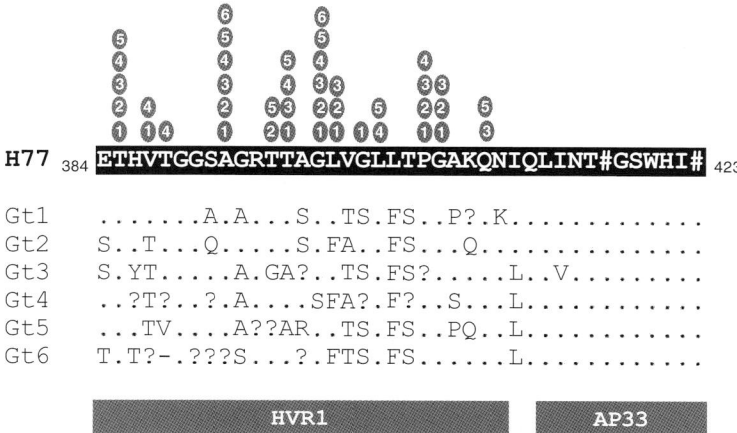

Fig. 2 Alignment of genotype consensus amino acid sequences relative to the H77 reference strain, showing distribution of amino acid sites identified as under diversifying selection in E1E2 sequences relative to their homologous positions in the H77 reference strain polyprotein (accession no. NC_004102). Symbols *above* the sequence indicate positively selected sites for sequences representative of each genotype 1 through 6. Also shown is the region targeted by the MAb AP33. #, putative N-linked glycosylation site (NXT or NXS). ?, no overall consensus amino acid at this position. Modified from Brown et al. (2007)

study examining a series of E1E2 sequences in the form of retroviral pseudoparticles from samples collected over a 26-year period of one patient found that sequential serum antibodies obtained at the same time points of the E1E2 sequences either fail or showed reduced Vn activity demonstrating a continuous generation of escape mutants predominately in HVR1 (von Hahn et al. 2007). Despite this caveat, studies have shown that positive selection is not evenly distributed across HVR1 and instead is focused on a few discontinuous residues (Brown et al. 2007; McAllister et al. 1998; Penin et al. 2001). By using consensus HVR1 sequence mimotopes, Puntoriero et al. (1998) were able to induce cross-reactive antibody responses in immunized mice. Therefore, isolation of antibodies focused on the conserved HVR1 residues may still have therapeutic promise.

We recently performed in-depth analysis of the E1E2 evolution, both across diverse genotypes as well as within individual-patient quasispecies (Brown et al. 2005, 2007), and this work has shown that much of the E1E2 coding region is under very strong purifying selective pressure. This is particularly evident in regions of known or suspected functional importance. We have shown that several discontinuous regions of E2 contain highly conserved residues that are involved in binding to CD81, a receptor for HCV (Owsianka et al. 2006) and, importantly, some of these residues are targeted by Vn antibodies. For example, MAb AP33 (see Sect. 3.5, "Antibodies to Linear Epitopes" below) has a broadly Vn phenotype that can be attributed to the extreme conservation of its epitope and the importance of this region of E2 in CD81 binding (Owsianka et al. 2005, 2006; Tarr et al. 2006). Utilizing Vn antibodies that target such highly conserved, functionally constrained epitopes will limit the likelihood of viral escape, and therefore maximize Vn antibodies' therapeutic potential. Even if escape occurs within these conserved epitopes, it is probable that this will come at a replicative cost to the virus, such as decreased receptor binding. Such a relationship is not unprecedented; there is mounting evidence that mutations in and around the CD4 binding domain of gp120 are associated with reduced sensitivity to Vn antibodies but at the cost of efficient CD4 receptor binding (Beaumont et al. 2004; Pinter et al. 2004; Pugach et al. 2004).

3 Humoral Immunity

What is the evidence that Vn antibodies can be protective? Until recently, the role of antibodies in preventing and controlling HCV infection has been less defined because of difficulties in having efficient and reliable in vitro systems to grow HCV. Early clinical trials with IgG therapy achieved significant prophylactic effects on transfusion-associated HCV hepatitis (Sugg et al. 1985; Sanchez-Quijano et al. 1988) and a reduction of infection among sex partners of HCV-infected patients receiving gamma-globulin (Piazza et al. 1997). Experimental animal studies have shown that IgG therapy delayed the onset of acute infection in HCV-challenged chimpanzees (Krawczynski et al. 1996). If the challenge HCV inoculum is first pre-incubated with HCV-specific Ig plasma, complete protection can be achieved (Farci et al. 1994). This led to studies showing that serum antibodies from patients with chronic HCV are directed at HVR1 (Shimizu et al. 1994, 1996; Zibert et al. 1995; Rosa et al. 1996). Additional findings were later obtained in chimpanzees by pre-incubating the infectious plasma with a rabbit antiserum specific for HVR1 (Cooreman and Schoondermark-Van de Ven 1996; Farci et al. 1996). Vaccination with recombinant HCV E1 and E2 proteins also provided some protection (Choo et al. 1994). Supporting the view that the humoral immune response is at least partly protective is the finding that circulating HCV virions in a chronically infected chimpanzee exist in two forms, as free virions and as immune complexes, with the latter displaying lower infectivity (Hijikata et al. 1993).

Acute infection with HCV is usually silent, with most infected individuals going on to develop persistent infections (Alter and Seeff 2000). Nonetheless, 20%–25% of acutely infected individuals clear viremia without disease progression. Also, a study of injecting drug users (IDU) comparing previously HCV-infected individuals (HCV antibody-positive and HCV RNA-negative) and those without evidence of prior infection (HCV antibody-negative and RNA-negative) found previously infected IDU were 12 times less likely than those with first time exposure to develop persistent infection (Mehta et al. 2002). Peak viral load on average was $10^{-1.8}$ lower in previously infected IDU. These findings indicate that adaptive immunity can prevent disease progression. Understanding the mechanisms contributing to acute infection resolution and control of disease persistence will be essential for vaccine and immunotherapeutic development. Conversely, identifying factors contributing to virus escape will help define those protective determinants most likely to give broad protection. In a recent study of a single-source outbreak of hepatitis C in a cohort of women, viral clearance was found to be associated with a rapid induction of Vn antibodies in the early phase of infection, the amounts of which were reduced or lost altogether following recovery from virus infection. In contrast, chronic HCV infection in this cohort was characterized by absent or low-titer Vn antibodies in the early phase and during persistence of infection despite the induction of cross-neutralizing antibodies in the late phase of infection. These data suggest that rapid induction of Vn antibodies during the early phase of infection may contribute to control of HCV infection (Pestka et al. 2006).

Vn epitopes can be broadly classified as either isolate-specific or broadly cross-reactive. The N-terminus of E2, encompassing HVR1 and the region immediately downstream contains potent linear epitopes (Flint et al. 2000; Flint et al. 1999; Owsianka et al. 2001; Triyatni et al. 2002; Zibert et al. 1997). Conservation in the region downstream of HVR1 leads to some cross-reactivity, although high levels of diversity seen between different genotypes can abrogate Vn (Bartosch et al. 2003a). More broadly, Vn antibodies have been demonstrated (Logvinoff et al. 2004) and these are usually directed against conformational epitopes within E2 (Allander et al. 2000; Bugli et al. 2001; Habersetzer et al. 1998; Hadlock et al. 2000; Ishii et al. 1998). E1-specific Vn antibodies are rare (reviewed in Depraetere and Leroux-Roels 1999), probably because this protein has low immunogenicity. However, there is evidence that E1-specific responses can be invoked following experimental vaccination and these responses may be protective (Rollier et al. 2004). Further, E1-specific Vn antibodies are present in some HCV-infected patients. We have previously described the use of an HCV-producing lymphoma cell line (Sung et al. 2003) to confirm Vn activity of a human monoclonal antibody (HMAb), H-111, to an epitope located on the N-terminal end of E1 that also blocks baculovirus-expressed HCV-like particles binding to target cells (Baumert et al. 1998; Keck et al. 2004b). The presence of E1 Vn antibodies is supported by recent studies showing that E1-specific sera are capable of neutralization of both retroviral pseudoparticles carrying HCV E1E2 proteins as well as cell cultured HCV (Dreux et al. 2006). Different groups have suggested that both E1 and E2 contain putative fusion domains and the development of HMAbs to E1 should facilitate functional and structural analyses of this HCV envelope protein.

3.1   Human Antibodies to Conformational Epitopes

There are two fundamental approaches to characterize the antibody response through the generation of HMAbs: the immortalization of antigen-specific B cells and the combinatorial approach of cloning Ig gene repertoires; each with its distinct attributes. The first approach includes Epstein-Barr virus (EBV) transformation, which itself is of limited utility because of the inherent instability of antibody secretion in EBV-transformed B cells. This instability can be addressed by fusing EBV-activated B cells to a human heteromyeloma fusion partner (Foung et al. 1984). The refinements of cell fusion methods have led to successes in generating relevant HMAbs with as few as 10^5 human B cells from peripheral blood (Foung et al. 1984, 1990; Zimmermann et al. 1990; Perkins et al. 1991). The advantages of this approach are the relative ease in obtaining antigen-specific B cells from peripheral blood and the ability to more directly analyze the B cell response under disease constraints. EBV transformation or activation of peripheral B cells followed by fusion to a human heteromyeloma has been used successfully by a number of groups to isolate HCV-specific HMAbs (Siemoneit et al. 1995; Habersetzer et al. 1998; Hadlock et al. 2000; Keck et al. 2004b). The second approach involves the construction of combinatorial libraries expressing human Fab fragments on the surface of M13 phage (Burton and Barbas 1994). This approach has the theoretical ability to examine millions of Fabs and consequently the full display of possible antibody diversity in the human B cell repertoire. Enrichment of combinatorial libraries constructed from the bone marrow of HCV-infected individuals has led to the recovery of a number of recombinant HMAbs to HCV E1E2 glycoproteins (Allander et al. 2000; Burioni et al. 1998; Burioni et al. 2001; Schofield et al. 2005; Zhong et al. 2005).

The majority of antibodies capable of interacting with diverse strains of HCV are directed to conformational epitopes within E2 (Allander et al. 2000; Bugli et al. 2001; Habersetzer et al. 1998; Bartosch et al. 2003b; Hadlock et al. 2000). In a panel of HCV HMAbs, cross-competition studies showed that the E2 glycoprotein contains three immunogenic conformational domains, designated A, B, and C, and these are accessible on the surface of retrovirus particles incorporating HCV glycoproteins, HCVpp (Keck et al. 2004a). Without knowing the exact location of Vn and non-Vn HMAbs to conformational epitopes on E2, their spatial relationship was assessed based on competition studies (Keck et al. 2005). Each HMAb was biotinylated and binding was tested with competing HMAb. Relatedness of the HMAbs to each other was determined by a modified unweighted pair-group method using arithmetic averages (UPGMA) (Fitch and Margoliash 1967; Keck et al. 2004a; Sneath and Sokal 1973). This method assumes the extent of bidirectional inhibition as the extent of epitope overlap by the competing antibodies. Unidirectional inhibition or enhancement is interpreted as proximal but not overlapping epitopes. This approach spatially places paired antibodies with the highest bidirectional inhibition as most related and an example of such an analysis is shown in Fig. 3. The cluster analysis yielded three distinct immunogenic domains. The epitopes recognized by HMAbs CBH-4G, -4B, and -4D constitute domain A.

A second domain, B, is characterized by CBH-2, -5, -8C, -8E, and -11, and a third domain C by CBH-7 and -23. HMAbs within a domain have significant bidirectional inhibition and minimal competition with HMAbs in other domains. For recently isolated HMAbs (Fig. 3 in italics), their placements were based on cross-competition with representative antibodies from each domain. CBH-20, -21, and -22 belong to domain A, and CBH-23 to domain C. HMAbs to domain B and C inhibit E2-CD81 interaction and HMAbs to domain A do not (Hadlock et al. 2000).

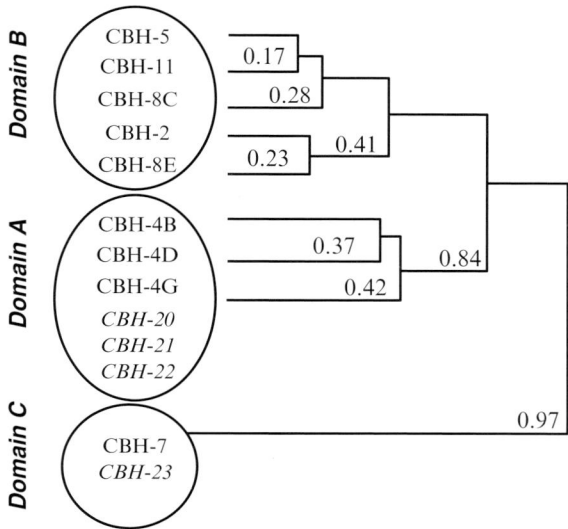

Fig. 3 Competition analysis of HCV HMAbs and proposed model for three immunogenic domains. Phylogenetic grouping of HCV HMAbs is based on cross-competition. *Solid lines with numbers* indicate the relatedness of the two adjoining antibodies, the smaller the number, the greater the relatedness. *Circles* are clusters of antibodies in a specific domain. Competition results used for calculation are the mean values obtained from 2–5 separate experiments (Keck et al. 2004a)

3.2   Virus Neutralization with HCV Retroviral Pseudoparticles

A robust means to analyze Vn potency and breadth is the infectious HCV retroviral pseudoparticle (HCVpp) system, whereby infection by retroviral particles containing an appropriate reporter gene is mediated by HCV E1E2 envelope proteins (Bartosch et al. 2003b; Bartosch et al. 2003c; Hsu et al. 2003). These HCVpp preferentially infect human hepatocytes and hepatocellular cell lines, and the E1E2 present in these particles has been shown to represent the functional and antigenic noncovalent E1E2 heterodimer formed by their recognition by a panel of conformation-dependent E2-specific HMAbs (Op De Beeck et al. 2004). All E2 HMAbs to conformational epitopes were able to immunoprecipitate HCVpp, suggesting their expression on the HCVpp surface (Keck et al. 2004a). Studies with HCVpp have provided important insights into the early stages of virus entry (Bartosch et al. 2005, 2003c; Goffard et al. 2005; Lavillette et al. 2005; Voisset et al. 2006). With genotype 1b HCVpp infection on Huh-7 cells, HCV E2 HMAbs in domains B (CBH-2, -5, -8C, -11) and C (CBH-7, -23) have Vn activities as measured by reduction of luciferase activity compared with the infection medium contains phosphate-buffered solution (PBS) (no antibody) or RO4 (Keck et al. 2004a, 2005). All domain A HMAbs were non-Vn (Fig. 4).

To assess the breadth of Vn, three HCV HMAbs with broad binding profiles to different genotypes, CBH-2, -5, and -7, were tested with six HCV genotype HCVpp (Fig. 5). CBH-2 showed Vn with one of two isolates of genotype 1a, genotypes 1b, 2a, 2b, 5, 6, and weak Vn with 4. CBH-5 showed Vn with 1a, 1b, 2a, 2b, 4, 6, and weak Vn with 5. CBH-7 showed Vn with only 1a, weak Vn with other genotypes, and no activity with one of two isolates of genotype 4. The breadth of Vn is roughly correlated with the breadth of binding with each Vn HMAb (A.M. Owsianka, A.W. Tarr, Z.Y. Keck, T.K. Li, J. Witteveldt, R. Adair, S.K.H. Foung, J.K. Ball, and A.H. Patel, manuscript in preparation).

The relationship between inhibiting E2-CD81 interaction and HCVpp neutralization also holds true with recombinant HMAbs to E2 (Schofield et al. 2005). Cross-competition studies with HMAbs from phage display libraries mapped a cluster of

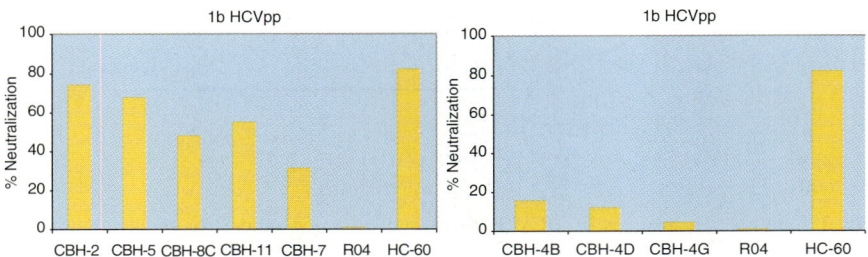

Fig. 4 Genotype 1b HCVpp Vn with E2 HMAbs. The HCVpp neutralization assay was performed as described by Keck et al. (2005) with each HMAb tested at 20 µg/ml. HC60, an HCV-positive human serum, was tested at 1:100 dilution. RO4 is an isotype-matched HMAb to CMV

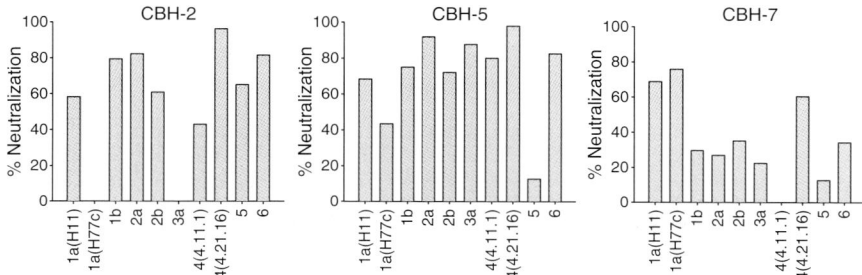

Fig. 5 Neutralization of six genotype HCVpp (ten isolates). Huh-7 cells were seeded at 8×10^3 cells per well in a 96-well plate 24 h before infection. HCVpp bearing envelopes from HCV genotype *1a* (*H11, H77c*), *1b* (1B5.23), *2a* (2A1.2), *2b* (2B2.8), *3a* (3A13.6), *4* (*4.11.1, 4.21.16*), *5* (5.15.7), and *6* (6.5.8) were incubated with HMAbs (*CBH*), an irrelevant isotype control HMAb RO4, or PBS (no MAb control) for 1 h at 37°C. After 15 h the HCVpp medium was replaced with fresh complete medium and incubated for an additional 72 h. Vn activity of an antibody was determined by the percentage reduction of luciferase activity compared with the infection medium containing PBS (not shown)

related epitopes inhibiting E2-CD81 interaction or neutralization of binding (NOB)-positive and three unrelated epitopes that are NOB-negative (Bugli et al. 2001).

3.3 Virus Neutralization with Cell Culture Infectious HCV Virions

Three groups recently developed full-length HCV RNA genomes that are able to replicate efficiently and produce infectious viral particles when transfected into human hepatoma cells (Huh-7) (Lindenbach et al. 2005; Wakita et al. 2005; Zhong et al. 2005). The availability of these and other cell-cultured infectious HCV virions (HCVcc) should greatly accelerate studies of HCV biology (Cai et al. 2005; Lindenbach et al. 2005; Pietschmann et al. 2006; Wakita et al. 2005; Yi et al. 2006; Zhong et al. 2005). Another group reported that stable human hepatoma cell lines containing a chromosomally integrated cDNA of HCV genotype 2a (JFH1) RNA constitutively secrete HCVcc into the medium (Cai et al. 2005). Employing this source of HCV virions, HMAbs to three distinct immunogenic domains on HCV E2 to neutralize HCVcc were tested (Keck et al. 2007). Vn was determined by measuring the levels of HCV NS3 protein expression by Western blot analysis. Figure 6 shows the inability of each domain A HMAb and the abilities of representative domain B (CBH-5) and C (CBH-7) HMAbs to neutralize HCV infectivity to Huh-7.5 cells. All domain A HMAbs had no effect on NS3 protein levels (Fig. 6A). HCVcc was completely neutralized at low antibody concentrations by all domain B HMAbs (as represented by CBH-5 in Fig. 2B and more fully discussed below). However, domain C HMAb, CBH-7, showed Vn activity only at 5 µg/ml IgG (Fig. 6B). The GAPDH

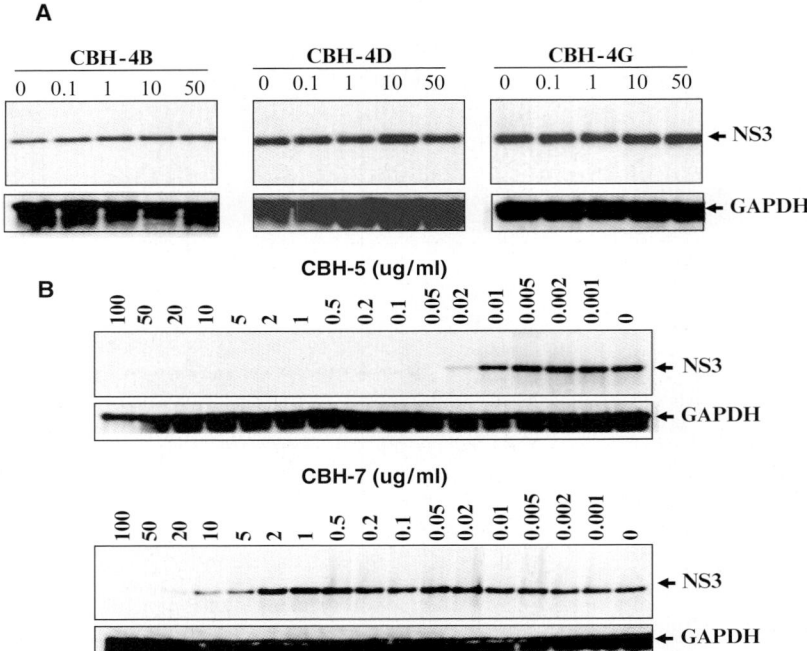

Fig. 6 Effect of HMAbs on genotype 2a HCVcc infectivity as determined by NS3 expression. HCVcc was incubated with each HMAb at increasing concentrations prior to infection to Huh-7.5 cells pre-seeded in a 24-well plate. At 3 h post infection (p.i.) the HCVcc/antibody-containing medium was removed, the cells washed twice with PBS followed by incubation at 37°C in fresh medium. Cells were harvested for Western blotting analysis at 72 h p.i. **A** Domain A HMAbs showed no reduction of NS3 expression. **B** Quantitative HCVcc Vn with *CBH-5* and *-7* shown as respective representative domain B and C antibodies. HCV NS3 protein expression was determined by a MAb specific to NS3 (Cai, Zhang et al. 2005). The GAPDH protein used as an internal control was detected using an anti-GAPDH MAb (Abcam)

protein used as an internal control was detected by using an anti-GAPDH MAb (in Fig. 6A and 6B; Abcam, Cambridge, MA). Similar findings of Vn neutralization with domain B and C HMAbs were obtained with infectious genotype 2b HCV virions from a B cell lymphoma-derived cell line that continuously produces infectious HCV virions in culture (Sung et al. 2003; M. Lai, personal communication). Collectively, these findings support an immunogenic model of HCVcc E2 containing three distinct functional domains as previously shown with HCVpp studies.

3.4   Virus Neutralization Potency with HCV Virions

A more quantitative measure of neutralization potency was determined by the inhibitory antibody concentration that reduces HCV infectivity 90%, IC_{90} (Keck et al. 2007).

Table 1 Neutralization titer

Antibodies	IC$_{50}$[a]		IC$_{90}$[a]	
	µg/ml	nmol/l	µg/ml	nmol/l
CBH-2	0.057	0.378	0.350	2.330
CBH-5	0.056	0.375	0.096	0.640
CBH-8C	0.205	1.367	0.980	6.530
CBH-11	0.283	1.886	4.061	27.070
CBH-7	25.580	170.530	49.800	332.000

[a]Antibody concentration to reach 50% and 90% viral neutralization

HCVcc neutralization conferred by each HMAb is summarized in Table 1. The IC$_{90}$ for HMAb CBH-5 was 0.01 µg/ml, while CBH-7 required approximately 50 µg/ml. The neutralization potency is in the order of CBH-5>CBH-2>CBH-8C>CBH-11>CBH-7. Another observation was that the HCVcc neutralization profiles for CBH-5 and CBH-7 displayed a linear relationship with antibody concentration from IC$_{50}$ to IC$_{90}$ (Fig. 6B and Table 1). However, the profiles for CBH-2, -8C, and -11 displayed a nonlinear relationship between IC$_{50}$ and IC$_{90}$, in which 5 to 14 times more antibodies were required to achieve IC$_{90}$ than IC$_{50}$. The possible contributions to differences with Vn profiles include the affinity of antibody binding to different viral epitopes, kinetics of antibody association and dissociation with targeted antigen, IgG subclass, and mechanisms of Vn whereby these antibodies inhibit different facets of HCVcc interactions with its putative receptor(s) (Harris et al. 1997). However, IgG subclass and antibody affinity are unlikely factors in this case as these antibodies are all IgG$_1$ and their relative K_d values are similar except for CBH-5 (Keck et al. 2007). More studies will be required to find out whether these antibodies block at distinct steps in virus–receptor interactions. In summary, the immunogenic organization of HCVcc E2 glycoprotein consists of three distinct domains, with Vn epitopes restricted to two domains as described for HCVpp. The fact that the ability to block E2 binding to CD81 is predictive of Vn provides additional support that CD81 is a required molecule for HCVpp and HCVcc entry.

3.5 Antibodies to Linear Epitopes

While Vn E1 antibodies have been described (Keck et al. 2004a; Dreux et al. 2006; Pietschmann et al. 2006), the majority of antibodies in natural infection are targeted to the E2 protein. Antibodies that demonstrate broad Vn capacity described to date generally tend to be directed against conformational epitopes within E2 (Habersetzer et al. 1998; Ishii et al. 1998; Allander et al. 2000; Hadlock et al. 2000; Bugli et al. 2001). However, Vn antibodies that recognize linear epitopes within HCV E2 have also been reported. Farci et al. (1996) first described a rabbit hyperimmune serum directed against HVR1 of E2 that was able to neutralize virus in vitro and thus protect chimpanzees in experimental infections. Subsequently, Zibert et al. (1995, 1997) showed that patient antibodies to HVR1 as well as a rabbit immune serum to that region were able to specifically block virus binding to cells. More recently, antibodies targeting this region have been shown to inhibit binding of E2 to cells (Scarselli et al. 2002), HCV virus-like particles (VLPs), and serum-derived HCV (Steinmann et al. 2004; Zhou et al. 2000; Barth et al. 2005), as well as HCVpp entry into target cells (Bartosch et al. 2003a; Hsu et al. 2003; Owsianka et al. 2005). The HVR1, which comprises the N-terminal 27 residues of E2 (aa 384–410), is highly variable among HCV genotypes as well as subtypes belonging to the same genotype. Deletion of HVR1 was shown to attenuate virus infectivity both in the chimpanzee model and also in HCVpp assay (Bartosch et al. 2003c; Forns et al. 2000). Together, these indicate an important role of this region in virus entry. However, due to the high variability of this epitope, these antibodies exhibit poor cross-neutralization potency across different HCV isolates. The use of conserved HVR1 mimotopes has been proposed to overcome problems of restricted specificity (Cerino et al. 2001; Roccasecca et al. 2001; Zucchelli et al. 2001).

Several antibodies targeting different linear regions (other than HVR1) within E2 that inhibit E2–CD81 interaction in in vitro assays have been identified. The first of these regions lies immediately downstream of the second hypervariable region between positions 480 and 493 (Flint et al. 1999), the second spans residues 528 to 535 (Owsianka et al. 2001; Clayton et al. 2002), and a third region encompasses residues 544 to 551 (Flint et al. 1999; Owsianka et al. 2001). In addition, antibodies targeting amino acids 412–423 and 432–447 are also capable of blocking CD81 binding (Owsianka et al. 2001; Hsu et al. 2003). Barth et al. (2003, 2006) recently showed that antibodies specific to regions 516–530 block the interaction between E2 and heparin sulfate, a candidate receptor for HCV, and they also inhibit HCVpp entry. Interestingly, however, those antibodies recognizing regions 480–493, 528–535, and 544–551 failed to neutralize HCVpp infection of target cells (Hsu et al. 2003). Furthermore, the region 432–447 displays a high degree of variation among different HCV isolates, and therefore the antibodies targeting this region have restricted cross-neutralizing potency (Owsianka et al. 2005).

The region 412–423, which is located immediately downstream of HVR1, has been shown to contain a potent neutralizing epitope (Clayton et al. 2002; Flint et al. 2000; Triyatni et al. 2002). This epitope, defined by the mouse MAb AP33 and a rat MAb 3/11, inhibited the interaction between CD81 and a range of presentations of E2, including soluble E2, E1E2, and virus-like particles (VLPs) (Owsianka et al. 2001). Moreover, MAb AP33 inhibited the interaction between E2 and heparin sulfate (Barth et al. 2006). Indeed, the MAb AP33 was reported to potently neutralize HCVpp-bearing envelope glycoproteins derived from all six HCV genotypes and major subtypes (Owsianka et al. 2005). As shown in Fig. 7 and Table 2, the IC_{50} of AP33 when neutralizing HCVpp of diverse genotypes ranged from approximately 0.6 up to 32 µg/ml. The rat MAb 3/11, which recognizes the same E2 region, also neutralized diverse HCVpp, albeit with a lower potency than MAb AP33 (Tarr et al. 2006). Similarly, both MAbs AP33 and 3/11 were capable of neutralizing the genotype 2a HCVcc in cell-culture infection assays, with the former more potent than the latter (Fig. 8). Fine mapping identified four highly conserved residues within the E2 region 412–423 crucial for MAb AP33 binding,

Fig. 7 MAb AP33-mediated neutralization of HCVpp-bearing envelope
glycoproteins from HCV genotypes
1a (*1A H77* and *1A14.36*), 1b
(*1B12.6*), 2a (*2A2.4*), 2b (*2B1.1*), 3a
(*3A13.6*), 4 (*4.21.16*), 5 (*5.15.11*),
and 6 (*6.5.340*). Taken from
Owsianka et al. (2005)

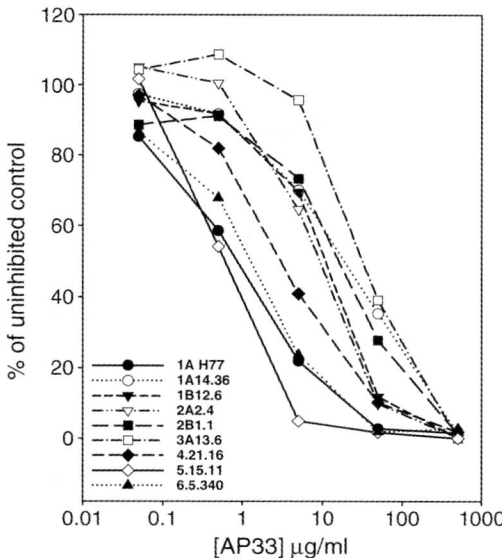

Table 2 MAb AP33 concentration required to achieve 50% (IC$_{50}$) or 90% (IC$_{90}$) inhibition of infection by diverse HCVpp shown in Fig. 7

Genotype	HCVpp isolate	IC$_{50}$	IC$_{90}$
1A	H77	5.9 nM (0.9 µg/ml)	145 nM (22.0 µg/ml)
1A	UKN1A.14.36	132.0 nM (20.0 µg/ml)	198 mmM (300.0 µg/ml)
1B	UKN1B.12.6	72.0 nM (12.0 µg/ml)	528 nM (80.0 µg/ml)
2A	UKN2A2.4	59.4 nM (9.0 µg/ml)	363 nM (55.0 µg/ml)
2B	UKN2B1.1	119.0 nM (18.0 µg/ml)	165 mM (250.0 µg/ml)
3A	UKN3A13.6	211.0 nM (32.0 µg/ml)	198 mM (300.0 µg/ml)
4	UKN4.21.16	19.8 nM (3.0 µg/ml)	396 nM (60.0 µg/ml)
5	UKN5.15.11	4.0 nM (0.6 µg/ml)	26.4 nM (4.0 µg/ml)
6	UKN6.5.340	8.6 nM (1.3 µg/ml)	145.2 nM (22.0 µg/ml)

whereas three residues were found to be critical for MAb 3/11, only two of which were shared with MAb AP33 (Tarr et al. 2006). Thus, this MAb targets an overlapping yet distinct epitope, indicating that this region harbors multiple Vn epitopes (Tarr et al. 2006).

Analysis of over 5,500 sequences obtained from the Genbank database showed that the AP33 epitope is highly conserved. Identifying potently Vn antibodies with epitopes conserved across all isolates of HCV is an essential step in the development of a successful vaccine. They could have a future role in the treatment of HCV infection and they might also serve to define future vaccine candidates. It will be important to define whether or not passive transfer of AP33 and 3/11 can protect in animal model-challenge experiments. Similarly, immunization and challenge studies

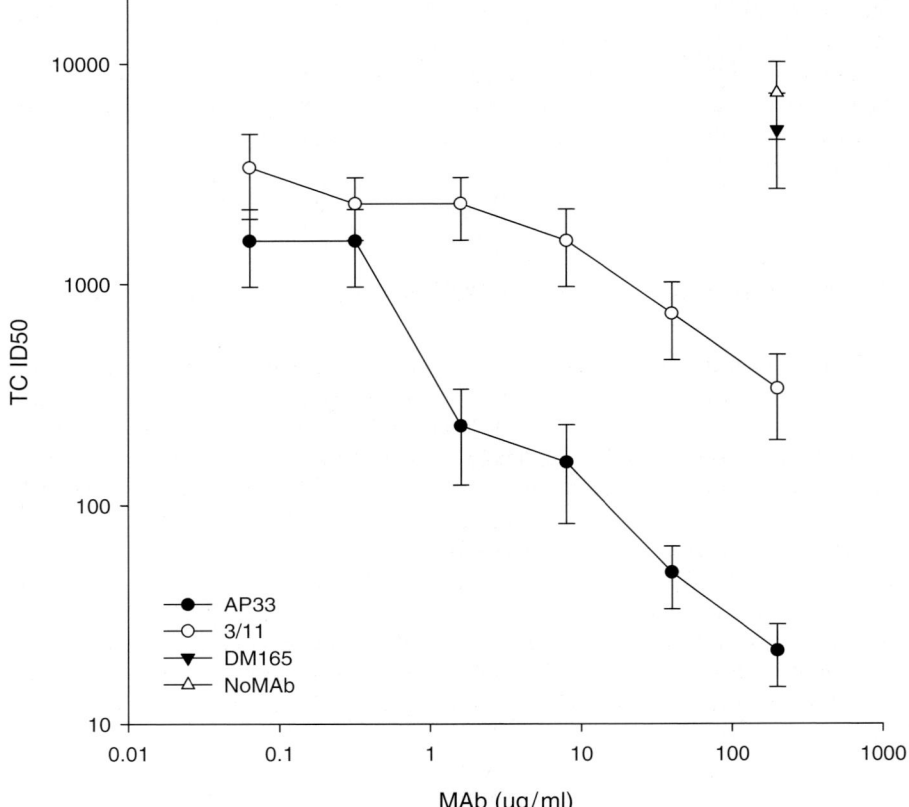

Fig. 8 Neutralization by MAbs AP33 and 3/11 of HCVcc infection of target cells in culture. HCVcc was pre-incubated with various concentrations of MAbs for 1 h at 37 °C. Serial dilution of each mixture was used to infect Huh-7 cells. Following incubation at 37 °C for 4 days the cells were fixed with methanol and analyzed by indirect immunofluorescence for the presence of viral protein NS5a. The wells of infected cells were scored for the presence or absence of fluorescing cells, and the virus infectivity was determined as $TCID_{50}$ (tissue culture infectious dose) essentially as described by Lindenbach et al. (2005). *DM165* denotes a irrelevant MAb control of the same isotype as AP33

using various immunogens containing the AP33 or 3/11 epitope will define its usefulness as a vaccine candidate. However, studies with HIV-1 have shown that focusing the immune response on epitopes recognized by broadly Vn antibodies is a significant challenge.

In this context, the finding that MAbs AP33 and 3/11 potently neutralize the entry of diverse HCVpp, and also HCVcc, is significant, particularly as their epitope is linear and highly conserved across different genotypes of HCV, and as such it offers significant hope for the development of a successful HCV vaccine. The exact mechanism of neutralization by these antibodies is unknown, although inhibition of CD81-binding or heparin sulfate-binding (or both) is the most likely (Owsianka et al. 2001; Barth et al. 2006). Thus, the design of future vaccine candidates based on these epitopes and AP33 (or 3/11)-like therapeutic antibodies will require a better understanding, at the molecular level of the antigen–antibody interaction. Only with correct presentation of an immunogen will a vaccine generate the desired immune responses.

4 Mechanisms of Virus Neutralization

HCV is a member of the family Flaviviridae and is composed of three structural proteins, capsid, two envelope proteins, E1 and E2, and six nonstructural proteins (Robertson et al. 1998; Lindenbach and Rice 2001; McLauchlan et al. 2002). Similar to other flaviviruses, virus entry is thought to be mediated by envelope proteins, which are responsible for virus attachment and receptor-mediated endocytosis where a low pH environment in the endosomes triggers conformational structural changes leading to virus envelope fusion with the endosomal membrane (Bartosch et al. 2003c; Hsu et al. 2003; Op De Beeck et al. 2004). Antibody binding to E1 or E2 epitopes on virion surface should then lead to Vn by a number of mechanisms (Smith 2001; Fig. 9).

First, HCV antibodies may mediate aggregation of virus particles, leading to a reduced number of infectious virions, the pentameric status of IgM therefore being associated with increased neutralization potency. Second, the prevailing view for some

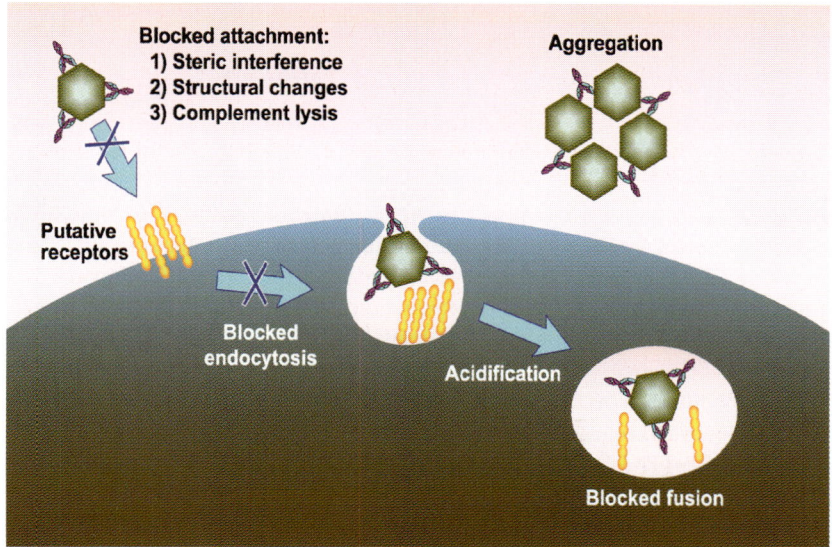

Fig. 9 Graphical representation of the mechanisms of virus neutralization

viruses is that Vn correlates with increased antibody binding to any virion surface site independent of the epitope recognized by the antibody. Vn is then the result of a critical number of binding sites being occupied, thus preventing virus entry through steric hindrance (Burton et al. 2001). Higher affinity antibodies will have a higher Vn while non-Vn antibodies either do not bind to virion surface or bind with low affinity. Alternatively, for some viruses, only specific surface epitopes that are involved in functional steps of virus entry have been proposed to be associated with neutralization (Houghton 1996). Steps targeted in this way include interaction with receptor and co-receptor, and initiation of viral envelope fusion with the cellular membrane. Our studies on HCVpp showed that E2 immunogenic domains A, B, and C are on the surface of these particles (Keck et al. 2004a). However, only domain B and C HMAbs mediate Vn, and domain A HMAbs are non-Vn, supporting the perspective that HCV virion attachment and entry are restricted to specific virion surface domains.

5 Mechanisms of Immune Escape

The mechanisms of viral escape include mutational escape from humoral (see Sect. 2, "Evolutionary Dynamics of HCV Envelope Genes") and cellular immunity, and viral infection of immune cells that leads to the induction of Ig hypermutation. The latter phenomenon is associated with a reduction in binding affinity and Vn activity of antibodies that are specific for HCV envelope proteins. Furthermore, antibody-dependent enhancement of infection and potential roles of lipoproteins and glycans to modify antibody binding to epitopes mediating Vn have been described (Voisset et al. 2006; Nielsen et al. 2006; Germi et al. 2002).

5.1 Mutational Escape from Cellular Immunity

The relationship between cell-mediated immunity and the outcome of HCV infection has been established by numerous studies. Memory CD8+ cytotoxic T cells (CTL) are required for protection against persistent HCV infection; at the same time, durable intrahepatic memory is likely established during acute HCV infection, since T cells recognizing HCV antigens have been recovered from the livers of chimpanzees several years after spontaneous clearance of infection (Shoukry et al. 2004). Consequently, the outcome of HCV infection may be dictated by escape mutations in the epitopes targeted by CTL (Erickson et al. 2001). On this issue, one report demonstrated that CTL escape mutations occurred in persistent HCV infection (Cox et al. 2005). A second report described that divergent and convergent virus evolution after a common-source outbreak of HCV affected disease outcome (Ray et al. 2005). Immune evasion leading to persistent infection, in contrast to recovery from viral infection, after acute HCV infection from a shared source has been reported (Tester et al. 2005), reinforcing the general relevance of this immune evasion mechanism to persistence of RNA viruses in humans (Bowen and Walker 2005a). Amino acid changes also can alter

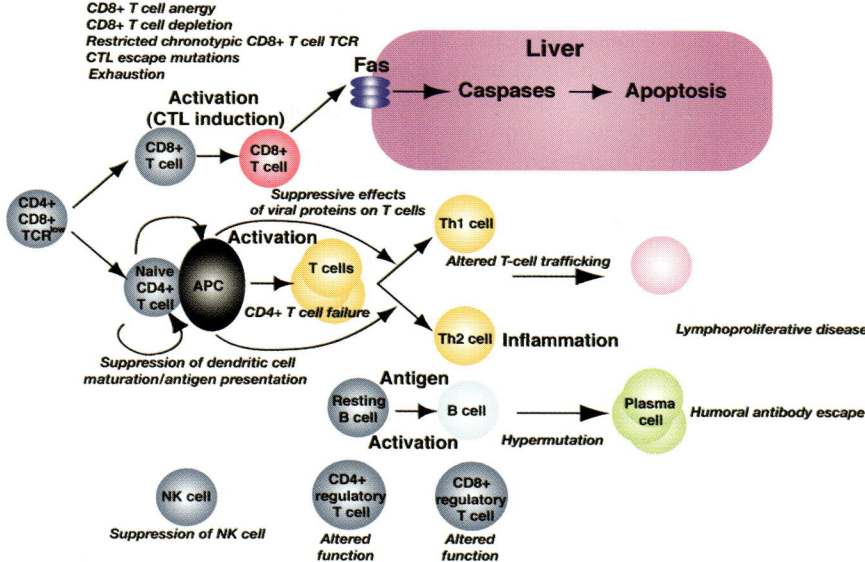

Fig. 10 Possible mechanisms of immune evasion by HCV infection or viral protein expression. All possible steps of immune dysfunction caused by HCV infection are *italicized*

CTL recognition of variant peptide–MHC complexes (Bowen and Walker 2005b). This and other mechanisms of HCV persistence are outlined in Fig. 10.

Successful immune responses in HCV infection generally target multiple major MHC class I-restricted epitopes in structural and nonstructural HCV proteins (Cooper et al. 1999; Shoukry et al. 2004). At the earliest time point studied in persons infected with HCV, highly activated CTL populations were observed that temporarily failed to secrete interferon (IFN)-γ, a "stunned" phenotype, from which they recovered as viremia declined (Lechner et al. 2000). In long-term HCV-seropositive persons, CTL responses were more common in those who had cleared viremia than those with persistent viremia, although the frequencies of HCV-specific CTL were lower than what was found in persons during and after resolution of acute HCV infection (Lechner et al. 2000).

CTL escape mutants are found during HCV infection in humans (Chang et al. 1997). Escape mutations in MHC class I-restricted epitopes are a feature of HCV infection that can diminish CTL responses via several mechanisms. For mutations in the CTL epitopes, marked fitness cost is not exacted by viral escape, and reversion to a more immunogenic ancestral state is not automatic upon passage to a host in which immune selection pressure is absent. It is tempting to speculate that this phenomenon might be due to low fitness cost associated with this particular mutation, thus allowing persistence of the variant sequence in the absence of immune selection pressure. A loss of epitope phenotype can also occur when amino acid anchor residues required for MHC binding are changed (Chang et al. 1997;

Erickson et al. 2001). Evidence for the emergence of escape mutations and their role in HCV infection are well-documented. The evolution of escape mutations in HCV is likely constrained by both intrinsic viral factors and certain characteristics of the adaptive immune response (see Sect. 2 above). There are several supporting reports on the lack of protective immunity against reinfection with HCV (Lai et al. 1994), the failure of naturally acquired antibodies to prevent reinfection of immune chimpanzees or humans (Farci et al. 1992), and emergence of CTL escape variants (Weiner et al. 1995).

Immune selection of HCV variants in humans includes the following steps (Bowen and Walker 2005a). First, mutations occur in immune epitopes of both structural and nonstructural viral proteins during acute infection. Second, evolution of quasispecies occurs in early infection. Third, a divergent or convergent CTL mutational evolutionary pattern from the prototypical epitope has been observed following a common-source HCV outbreak (Bowen and Walker 2005b; Ray et al. 2005; Tester et al. 2005). Some of these mutations are due to viral reversion to a more fit ancestral state (Cox et al. 2005; Ray et al. 2005). Nonetheless, these adaptive mutations can temper the effectiveness of CD8[+] CTL function (Franzin et al. 1995). There is a complex interplay between the breadth and specificity of the antigen-specific immune response and the degree to which mutations selected by this immune pressure govern viral reproduction. Fourth, limited epitope variation occurs during the late or chronic infection phase. Last, epitope variation can be present (restricting MHC allele vs nonrestricting MHC) without seroconversion during chronic infection (Post et al. 2004). Similarly in chimpanzees, immune selection of HCV variants includes the following observations (Cooper et al. 1999; Bowen and Walker 2005b). First, there is a minimal viral variation prior to onset of adaptive immune response. Second, escape mutations abrogating CTL function do occur in acute phase infection. Third, escape mutations evading the antiviral CTL response occurs in MHC class I-restricted epitopes (Weiner et al. 1995; Sasso 2000; Erickson et al. 2001).

Immune escape by mutations in CTL epitopes occurs by at least two mechanisms (Drummer et al. 2002; Bowen and Walker 2005a, b), the loss of T cell receptor (TCR) recognition (Ivanovski et al. 1998; De Re et al. 2000; Sasso 2000; inhibition of CTL response, antigenic sin, and preferential stimulation of response) and the loss of epitope by altered proteasome processing and reduced MHC class I binding. The amino acid substitutions within or adjacent to CTL epitopes can alter proteasomal processing, causing epitope destruction before transport to the endoplasmic reticulum for MHC binding (Seifert et al. 2004; Timm et al. 2004; Kimura et al. 2005). In HLA-A*01- and B*08-negative individuals, absence of these alleles was associated with evolution toward consensus within epitopes restricted by these MHC molecules (Ray et al. 2005).

The viral epitopes on the virus-specific CD4[+] and CD8[+] T cell frequently evolve during HCV infection, resulting in impaired effector function of HCV-specific CTL (Gruener et al. 2001; Wedemeyer et al. 2002) and a lack of protective immunity against reinfection with HCV (Lai et al. 1994). In the presence of the restricting allele, mutational escape of MHC class I-restricted epitopes may occur in four ways. First,

sustained cellular immune responses are associated with resolution of infection during the acute phase (Zuckerman et al. 1997). The diverse clonal CTL TCR repertoire due to sustained CD4+ T cell response may constrain the development of escape mutations in the restricting MHC molecule. Second, in the case of absent or weak CD4+ T cell responses, CTL responses are weak and CTL-escape mutations may not develop (Hanley et al. 1996; De Vita et al. 2000). Third, where the CD4+ T cell response fails, escape mutations might emerge, particularly if their fitness cost is low. Lastly, if the TCR repertoire of the clonal CTL is narrow in the absent or weak CD4+ T cell response, escape mutations may emerge (Grakoui et al. 2003; Shoukry et al. 2004). In the absence of the restricting allele, when there is low associated fitness cost or well-developed compensatory mutations and if there is no CTL-mediated immune pressure, the mutated sequence may persist or may mutate to an equally "fit" alternative. Where there is high fitness cost associated with the presence of an escape mutation and if there is no CTL-mediated immune pressure (minimal immune selection pressure), reversion to the wild-type ("fitter") sequence is likely to occur.

Epitope mutations in individual MHC alleles may alter the interaction of epitope with the immune system by at least four different mechanisms. First, mutational escape from CD8+ T cell immunity can occur (Bowen and Walker 2005a, b). Second, mutations in cognate epitopes in anchoring residues may lead to dissociation of the MHC–peptide complex. Third, mutations in the epitope or in flanking regions can alter proteasomal processing, leading to destruction of the epitope. Fourth, reduced TCR recognition of the neo-epitope–MHC complex is involved (Chang et al. 1997). TCR recognition may be reduced or possibly altered, leading to antagonism against responses to the wild-type peptide. Such mutated peptide–MHC complexes may alternatively antagonize responses to the wild-type epitope (Chang et al. 1997; Kaneko et al. 1997; Tsai et al. 1998). Nonsynonymous mutations may occur within CTL epitopes or within regions flanking these sequences.

Other immunological mechanisms (Bowen and Walker 2005b, c) are MHC class II-restricted escape mutations (Misiani et al. 1994), CD4+ T cell response from Th1 toward Th2 response (Casato et al. 2002), marked CD4+ CD25+ regulatory T cells (Rushbrook et al. 2005), regulatory CD8+ T cell, MHC class I-restricted antigen-specific regulatory activity with the potential to suppress antiviral T cells (Koziel et al. 1995), narrow CD8+ T cell receptor repertoire and impaired dendritic cell maturation in chronic hepatitis C patients (Auffermann-Gretzinger et al. 2001). Memory CD8+ T cells can vary in differentiation phenotype in different persistent virus infections (Appay et al. 2002). HCV persistence and immune evasion do occur in the absence of memory T cell help (Grakoui et al. 2003). A role of primary intrahepatic T cell activation in the "liver tolerance effect" has been reported (Bertolino et al. 2002). Finally, the upregulation of inhibitory receptor programmed death-1 (PD-1) expression has been shown to lead to HCV-specific CD8 exhaustion (Tseng and Klimpel 2002). Thus, due likely to interplay between the opposing forces of immune selection pressure and viral fitness cost, a variety of outcomes are possible upon initial infection with HCV, or after subsequent transmission of the virus to a recipient in whom the initial MHC class I alleles are not expressed.

5.2   HCV Infection of B Cells Induces Ig Hypermutation, Altering B Cell Immunity

Besides chronic hepatitis, liver cirrhosis, and hepatocellular carcinoma, HCV infection is also frequently associated with B lymphocyte proliferative disorders, including mixed cryoglobulinemia, a disorder characterized by oligoclonal proliferation of B cells, and non-Hodgkin's B cell lymphoma (Silvestri et al. 1997). These B cell diseases may be the result of infection of B cells by HCV or the activation of B cells by HCV envelope proteins. The HCV envelope proteins E1 and E2 are type I transmembrane proteins, with N-terminal ectodomains and C-terminal hydrophobic anchors. It has been suggested that HCV infects human cells through the interaction of E2 with a tetraspanin molecule CD81. CD81 is thought to be a cellular receptor for HCV, based on its ability to bind E2 (Pileri et al. 1998; Hsu et al. 2003; McKeating et al. 2004; Zhang et al. 2004). CD81 is a member of the tetraspanin family and is a component of the multimeric B cell antigen receptor complex (Levy et al. 1998). It is associated with other membrane proteins, which vary in different B cell lineages and include the signaling molecule CD19, complement receptor 2 (CD21), and interferon-inducible Leu-13 (CD225) protein (Takahashi et al. 1990; Levy et al. 1998). Binding of CD81 with E2 or certain MAbs against CD81 induces B cell aggregation, inhibits Daudi cell proliferation (Flint et al. 1999), stimulates T cells (Soldaini et al. 2003), and inhibits natural killer cell functions (Crotta et al. 2002; Tseng and Klimpel 2002). In addition, triggering of the CD81 signaling pathway in B cells enhances the production of tumor necrosis factor-α (TNF-α) (Altomonte et al. 1996). Correspondingly, HCV infection of primary macrophages has been reported to induce TNF-α production (Radkowski et al. 2004). Coengagement of the CD19–CD21–CD81 complex and the B cell antigen receptor lowers the B cell activation threshold by antigen-presenting cells or lipopolysaccharide (Carter and Fearon 1992). Lymphocytes in mice lacking CD81 develop normally but have altered proliferative responses and are deficient in antibody production, suggesting that CD81 is one of the essential receptors for the production of antibodies (Miyazaki et al. 1997). These observations suggest that HCV may modify the B cell receptor-associated signaling pathway by binding to CD81 or by infecting B cells.

Evidence has been presented that at least certain strains of HCV can infect and replicate in B cells (Sung et al. 2003; Machida et al. 2004b, 2005). HCV infection of B cells triggers double-strand DNA breaks in many cellular genes (Machida et al. 2004a, b). Interestingly, the mere engagement of B cells by purified E2 alone, without viral replication, induces double-strand DNA breaks specifically in the variable region of the Ig gene locus, leading to hypermutation in Ig of B cells (Machida et al. 2005). Other gene loci are not affected by the E2–CD81 interaction. Preincubation with the anti-CD81 MAb blocks this effect. E2–CD81 interaction on B cells triggers the enhanced expression of activation-induced cytidine deaminase (AID) and also stimulates the production of TNF-α (Machida et al. 2005).

Knockdown of AID by the specific small interfering RNA (siRNA) blocked the E2-induced double-strand DNA breaks (DSBs) and the hypermutation of the Ig gene (Machida et al. 2005). Therefore, HCV infection, through the E2–CD81 interaction, may modulate host's innate or adaptive immune response by activation of AID and ensuing hypermutation of immunoglobulin gene in B cells. These effects may contribute to HCV persistence and B cell lymphoproliferative diseases.

In addition to the increased mutation frequency in the Ig gene, HCV infection also induces hypermutation of many other cellular genes, including p53 genes (Machida et al. 2004b). Subsequent studies showed that the HCV-induced mutations of somatic genes, such as *p53*, are mediated by nitric oxide (NO) and reactive oxygen species (ROS) (Machida et al. 2004a). In contrast, as stated above, the Ig mutations are mediated by AID activation through the binding of HCV E2 protein to CD81. AID is involved in both the somatic hypermutation and class-switching recombination of the Ig gene in normal B cell development; it causes deamination of deoxycytidine to deoxyuracil (dU) in the template DNA strand, with preference for certain hot-spot motifs (Petersen-Mahrt et al. 2002). The resulting dU/dG pairs can be resolved by the mismatch repair system (Papavasiliou and Schatz 2002), uracil glycosylase endonuclease (Di Noia and Neuberger 2002), and error-prone DNA polymerases (Radkowski et al. 2004). Among the latter, Pol ι, Pol η, and Pol ζ are involved in these pathways (Zan et al. 2001; Zeng et al. 2001; Faili et al. 2002). Interestingly, AID, Pol ζ, and Pol ι are induced in HCV-infected B cells (Machida et al. 2004b), at least partially caused by the binding of HCV E2 protein to CD81, which, in turn, triggers DSBs and subsequent Ig hypermutations (Fig. 11).

Furthermore, this interaction induces TNF-α production by B cells. These effects have been confirmed in the natural HCV infection of B cell in vitro and in vivo. These findings implicate that, even in the absence of virus replication, the very act of virus binding to B cells can contribute to the pathogenesis of HCV by Ig hypermutation and TNF-α production.

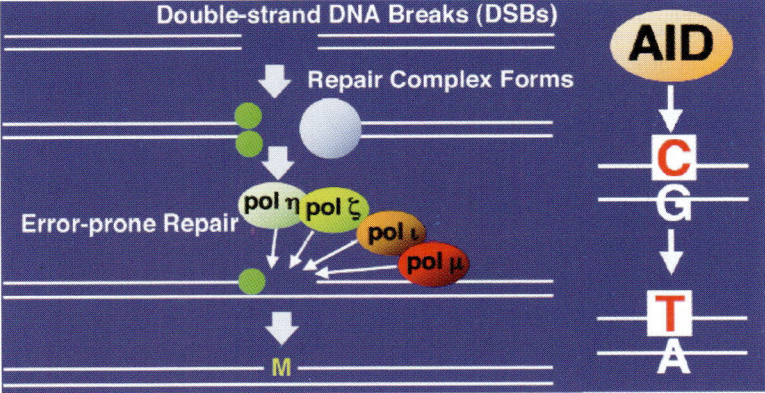

Fig. 11 Postulated mechanism of Ig hypermutation

The E2–CD81 interaction enhances mutation frequencies specifically in the Ig gene, but not in other genes, such as *p53*. The basis for such a specific effect is still not completely clear. E2–CD81 interaction may trigger a signaling response similar to that triggered by anti-CD40, interleukin (IL)-4, and other cytokines in B cells (Muramatsu et al. 1999; Chaudhuri et al. 2003). The normal somatic hypermutation mechanism of the Ig gene in B cells typically affects the genomic sequences within approx. 2 kb downstream from the transcription initiation site of the Ig gene (Rada and Milstein 2001), under the influence of the Ig gene enhancer (Goyenechea et al. 1997). The HCV E2-induced Ig hypermutation mirrors precisely this pattern, indicating that it is the result of hyper-activity of the normal Ig mutation mechanism. This specificity may explain the differential effects of E2–CD81 interactions on the Ig and p53 genes. E2 likely will bind most cell types, since CD81 is expressed ubiquitously. However, although E2 can bind to hepatocyte cell lines expressing CD81, it does not induce enhancement of expression of AID or DSBs in this non-B cell line (Machida et al. 2005). These findings suggest that the other components of the CD81 complex, including CD21 and CD19, are important for the signal transduction involved in the induction of AID (Fig. 12). Several protein kinases have been shown to associate with CD19 and CD21 but not with CD81 (Fearon and Carter 1995).

TNF-α is one of the earliest host responses to viral infections (Guidotti and Chisari 2001); it is an inflammatory cytokine, which can contribute directly or indirectly to viral pathogenesis. On the other hand, it may purge viruses from infected cells noncytolytically and mediate intracellular signaling by adjusting the redox potential of the cell (Wong and Goeddel 1988; Kizaki et al. 1993). Thus, the E2–CD81 interaction may contribute to HCV pathogenesis through TNF-α production. The activation of AID and subsequent Ig mutations may also contribute to the development of B cell lymphoma.

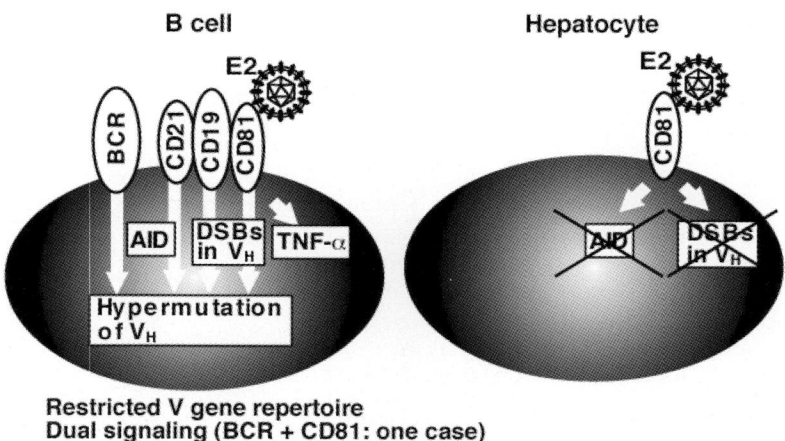

Fig. 12 A postulated signaling pathway for induction of hypermutation of the Ig gene in B cells or lack of induction in hepatocytes by E2 binding. *BCR*, B-cell receptor complex. A dual signaling model for B cell activation by HCV antigens has been reported (Weng and Levy 2003)

Another consequence of Ig hypermutation is that the specificity and avidity of antibodies produced from B cells will likely diverge following HCV infection. If the B cells producing HCV-specific antibodies are infected, it is predicted that the HCV-specific antibodies produced from these cells may lose their avidity and specificity within a short period of time. As a consequence, the neutralization activity or the antibody-mediated cell lysing activity may decline following HCV infection. This effect will enable the virus to escape from humoral immunity. This prediction has recently been proved correct by showing that HCV can infect human hybridoma cells producing E2-specific HMAbs, causing the decline of the avidity and specificity of the HCV-specific antibodies (K. Machida, Y. Kondo, J. Hwang, Y.C. Chen, K.T. Cheng, V.M. Sung, Z. Keck, S. Foung, J. Dubuisson, and M.C. Lai, unpublished observation). This mechanism will contribute to immune escape of HCV. Similarly, at least some strains of HCV can infect certain T cells, affecting T cell functions, including IFN-γ signaling (Kondo et al. 2007). The suppression of T cell functions may contribute to T cell anergy and immune escape of HCV.

5.3 Antibody-Dependent Enhancement of Infection

Complement-mediated enhancement of antibody function for neutralization of pseudotype virus containing HCV E2 chimeric glycoprotein has been reported (Meyer et al. 2002). Significant increases in the neutralization titers of E2 HMAbs and rabbit antiserum to HVR1 mimotopes have been observed upon addition of guinea pig complement. Complement activation occurred primarily by the classical pathway, since a deficiency in the C4 component led to a significant decrease in the level of Vn. During infection, HCV E2 glycoprotein induces a weak Vn antibody response; these antibodies can be measured in vitro by the surrogate pseudotype virus plaque reduction assay, and the neutralization function can be augmented by complement (Meyer et al. 2002). Therefore, it is possible that complement activation enhances infection, and lipoproteins and glycans can modify antibody binding to epitopes mediating Vn.

5.4 Potential Roles of Lipoproteins and Glycans to Modify Antibody Binding to Epitopes Mediating Virus Neutralization

Lipoproteins, high-density lipoproteins (HDL), low-density lipoproteins (LDL), and very low-density lipoproteins (VLDL) could potentially reduce the neutralizing effect of the HCV-specific antibodies by promoting HCV entry (Voisset et al. 2006). In studies on HCVpp, HDL inhibits HCV Vn antibodies by stimulating cell entry via activation of the scavenger receptor BI (SR-B1). SR-B1 mediates lipid transfer and is proposed as a cell entry cofactor of HCV (Dreux et al. 2006). HDL interaction with the SR-BI has been shown to strongly reduce Vn of HCVpp by Vn MAbs and HMAbs (Dreux et al. 2006). For serum and liver HCV virions, it is probable that VLDL/LDL are involved in virus entry (Nielsen et al. 2006). The majority of serum and liver HCV RNA can be precipitated by antibodies to apolipoprotein B and apolipoprotein E. As to the site of lipoprotein linkage to virions, a study on intracellular infectious HCVcc and extracellular infectious HCVcc suggests that acquisition of lipoproteins might occur during viral secretion from infected cells (Gastaminza et al. 2006). While it is possible that some Vn E2 epitopes are masked by lipoproteins as suggested by HCVpp studies, some HCVcc E2 epitopes remain exposed as shown by Vn HMAbs to domain B (Keck et al. 2007). In addition, some E2 HMAbs will immunoprecipitate serum and liver virions (Kumar et al. 1994). Antibodies targeted to E1 protein of HCV efficiently neutralize HCVpp and HCVcc in the presence of human serum. Functional features of HCV glycoproteins for pseudotype virus entry into mammalian cells have been reported (Meyer et al. 2000; Beyene et al. 2002). LDL receptor-related molecules partially inhibit E1 pseudotype virus infectivity, while CD81-related molecules interfere with E2 pseudotype virus infectivity. A further understanding of HCV entry and strategies appropriate for mimicking cell surface molecules may help in the development of new therapeutic modalities against HCV infection.

6   Final Remarks

Vn monoclonal antibodies will likely succeed in the prevention of HCV reinfection after liver transplantation. However, it should be noted that a small trial was recently performed where liver transplant recipients with HCV-associated liver failure received multiple infusions of high-dose immunoglobulin of 200 mg/kg per dose over 14 weeks, but no viral load decrease was detected (Davis et al. 2005). However, this study should be considered with caution. In addition to the concern that the dosage used was based on the protective amount previously observed in chimpanzees (Krawczynski et al. 1996), there are a number of other issues. First, specific HCV antibodies are only a small fraction of the total Ig and of these only a smaller fraction will be directed to the important targets involved in entry, i.e., the E1E2 envelope glycoproteins. Second, not all antibodies to E1 or E2 mediate Vn (Keck et al. 2004a). Both Vn and non-Vn HMAbs to HCV E2 have been described. The existence of non-Vn HMAbs to HCV E2 has similarly been observed in sera of patients with chronic HCV infection (Burioni et al. 2004). As chronically infected patients were probably the primary source of IgG used in this trial, it is probable that large portions of these antibodies were non-Vn. Finally, the viral load in the patients enlisted for this clinical trial was many logs higher than in the challenged animals, and the supposedly "high" dose thus may still be inadequate. Supporting this possibility is the observation that liver tissue viral load in biopsies tended to be lower in high-dose Ig recipients compared to low-dose recipients (Davis et al. 2005).

The studies summarized in this review support the feasibility of isolating Vn antibodies to broadly conserved epitopes among different HCV genotypes. While candidate Vn antibodies to highly conserved conformational and linear epitopes have been identified, a greater understanding of the factors contributing to virus escape will be required to help define those protective determinants most likely to give broad protection. The emergence of the escape viral mutants may be slowed by the combination of Vn antibodies and antivirals. In addition, the antibodies used should be broadly reactive to different HCV genotypes, each inhibiting at different steps of virus entry, and be synergistic in their ability to control virus infection. At least two antibodies to different epitopes are needed to minimize the concern of escape mutants. Additional studies will also be required on the possible role of lipo-proteins in masking virion surface domains involved in virus entry, although current data suggest that a specific cluster of epitopes (designated domain B) on HCV E2 remains exposed on low-density cell culture infectious HCV virions. In summary, the Vn antibodies clearly merit consideration as a therapeutic and preventative strategy against HCV reinfection in the liver transplantation setting.

Acknowledgements We thank Dr. Richard Brown for providing Figure 1. This work was supported in part by National Institutes of Health grants HL079381 and AI47355 to SKHF, EU contracts QLK2-CT-2001–01120 and MRTN-CT-2006–035599 to JB, and the Medical Research Council, UK to AHP.

References

Allander T, Drakenberg K, Beyene A, Rosa D, Abrignani S, Houghton M, Widell A, Grillner L, Persson MA (2000) Recombinant human monoclonal antibodies against different conformational epitopes of the E2 envelope glycoprotein of hepatitis C virus that inhibit its interaction with CD81. J Gen Virol 81:2451–2459

Alter HJ, Seeff LB (2000) Recovery, persistence, and sequelae in hepatitis C virus infection: a perspective on long-term outcome. Semin Liver Dis 20:17–35

Altomonte M, Montagner R, Pucillo C, Maio M (1996) Triggering of target of an antiproliferative antibody-1 (TAPA-1/CD81) up-regulates the release of tumour necrosis factor-alpha by the EBV-B lymphoblastoid cell line JY. Scand J Immunol 43:367–373

Appay V, Dunbar PR, Callan M, Klenerman P, Gillespie GM, Papagno L, Ogg GS, King A, Lechner F, Spina CA, Little S, Havlir DV, Richman DD, Gruener N, Pape G, Waters A, Easterbrook P, Salio M, Cerundolo V, McMichael AJ, Rowland-Jones SL (2002) Memory CD8+ T cells vary in differentiation phenotype in different persistent virus infections. Nat Med 8:379–385

Auffermann-Gretzinger S, Keeffe EB, Levy S (2001) Impaired dendritic cell maturation in patients with chronic, but not resolved, hepatitis C virus infection. Blood 97:3171–3176

Barth H, Schafer C, Adah MI, Zhang F, Linhardt RJ, Toyoda H, Kinoshita-Toyoda A, Toida T, Van Kuppevelt TH, Depla E, Von Weizsacker F, Blum HE, Baumert TF (2003) Cellular binding of hepatitis C virus envelope glycoprotein E2 requires cell surface heparan sulfate. J Biol Chem 278:41003–41012

Barth H, Ulsenheimer A, Pape GR, Diepolder HM, Hoffmann M, Neumann-Haefelin C, Thimme R, Henneke P, Klein R, Paranhos-Baccala G, Depla E, Liang TJ, Blum HE, Baumert TF (2005) Uptake and presentation of hepatitis C virus-like particles by human dendritic cells. Blood 105:3605–3614

Barth H, Schnober EK, Zhang F, Linhardt RJ, Depla E, Boson B, Cosset FL, Patel AH, Blum HE, Baumert TF (2006) Viral and cellular determinants of the hepatitis C virus envelope-heparan sulfate interaction. J Virol 80:10579–10590

Bartosch B, Bukh J, Meunier JC, Granier C, Engle RE, Blackwelder WC, Emerson SU, Cosset FL, Purcell RH (2003a) In vitro assay for neutralizing antibody to hepatitis C virus: evidence for broadly conserved neutralization epitopes. Proc Natl Acad Sci U S A 100:14199–14204

Bartosch B, Dubuisson J, Cosset FL (2003b) Infectious hepatitis C virus pseudo-particles containing functional E1-E2 envelope protein complexes. J Exp Med 197:633–642

Bartosch B, Vitelli A, Granier C, Goujon C, Dubuisson J, Pascale S, Scarselli E, Cortese R, Nicosia A, Cosset FL (2003c) Cell entry of hepatitis C virus requires a set of co-receptors that include the CD81 tetraspanin and the SR-B1 scavenger receptor. J Biol Chem 278: 41624–41630

Bartosch B, Verney G, Dreux M, Donot P, Morice Y, Penin F, Pawlotsky JM, Lavillette D, Cosset FL (2005) An interplay between hypervariable region 1 of the hepatitis C virus E2 glycoprotein, the scavenger receptor BI, and high-density lipoprotein promotes both enhancement of infection and protection against neutralizing antibodies. J Virol 79:8217–8229

Baumert TF, Ito S, Wong DT, Liang TJ (1998) Hepatitis C virus structural proteins assemble into viruslike particles in insect cells. J Virol 72:3827–3836

Beaumont T, Quakkelaar E, van Nuenen A, Pantophlet R, Schuitemaker H (2004) Increased sensitivity to CD4 binding site-directed neutralization following in vitro propagation on primary lymphocytes of a neutralization-resistant human immunodeficiency virus IIIB strain isolated from an accidentally infected laboratory worker. J Virol 78:5651–5657

Bertolino P, McCaughan GW, Bowen DG (2002) Role of primary intrahepatic T-cell activation in the 'liver tolerance effect'. Immunol Cell Biol 80:84–92

Beyene A, Basu A, Meyer K, Ray R (2002) Hepatitis C virus envelope glycoproteins and potential for vaccine development. Vox Sang 83 [Suppl 1]:27–32

Biggins SW, Terrault NA (2005) Treatment of recurrent hepatitis C after liver transplantation. Clin Liver Dis 9:505–523, ix

Booth JC, Kumar U, Webster D, Monjardino J, Thomas HC (1998) Comparison of the rate of sequence variation in the hypervariable region of E2/NS1 region of hepatitis C virus in normal and hypogammaglobulinemic patients. Hepatology 27:223–227

Bowen DG, Walker CM (2005a) Adaptive immune responses in acute and chronic hepatitis C virus infection. Nature 436:946–952

Bowen DG, Walker CM (2005b) Mutational escape from CD8+ T cell immunity: HCV evolution, from chimpanzees to man. J Exp Med 201:1709–1714

Bowen DG, Walker CM (2005c) The origin of quasispecies: cause or consequence of chronic hepatitis C viral infection? J Hepatol 42:408–417

Brown RJ, Juttla VS, Tarr AW, Finnis R, Irving WL, Hemsley S, Flower DR, Borrow P, Ball JK (2005) Evolutionary dynamics of hepatitis C virus envelope genes during chronic infection. J Gen Virol 86:1931–1942

Brown RJ, Tarr AW, McClure CP, Juttla VS, Tagiuri N, Irving WL, Ball JK (2007) Cross-genotype characterization of genetic diversity and molecular adaptation in hepatitis C virus envelope glycoprotein genes. J Gen Virol 88:458–469

Bugli F, Mancini N, Kang CY, Di Campli C, Grieco A, Manzin A, Gabrielli A, Gasbarrini A, Fadda G, Varaldo PE, Clementi M, Burioni R (2001) Mapping B-cell epitopes of hepatitis C virus E2 glycoprotein using human monoclonal antibodies from phage display libraries. J Virol 75:9986–9990

Bukh J, Miller RH, Purcell RH (1995) Genetic heterogeneity of hepatitis C virus: quasispecies and genotypes. Semin Liver Dis 15:41–63

Burioni R, Plaisant P, Manzin A, Rosa D, Delli Carri V, Bugli F, Solforosi L, Abrignani S, Varaldo PE, Fadda G, Clementi M (1998) Dissection of human humoral immune response against hepatitis C virus E2 glycoprotein by repertoire cloning and generation of recombinant Fab fragments. Hepatology 28:810–814

Burioni R, Bugli F, Mancini N, Rosa D, Di Campli C, Moroncini G, Manzin A, Abrignani S, Varaldo PE, Clementi M, Fadda G (2001) Nonneutralizing human antibody fragments against hepatitis C virus E2 glycoprotein modulate neutralization of binding activity of human recombinant Fabs. Virology 288:29–35

Burioni R, Mancini N, Carletti S, Perotti M, Grieco A, Canducci F, Varaldo PE, Clementi M (2004) Cross-reactive pseudovirus-neutralizing anti-envelope antibodies coexist with antibodies devoid of such activity in persistent hepatitis C virus infection. Virology 327:242–248

Burton DR, Barbas CF 3rd (1994) Human antibodies from combinatorial libraries. Adv Immunol 57:191–280

Burton DR, Saphire EO, Parren PW (2001) A model for neutralization of viruses based on antibody coating of the virion surface. Curr Top Microbiol Immunol 260:109–143

Cai Z, Zhang C, Chang KS, Jiang J, Ahn BC, Wakita T, Liang TJ, Luo G (2005) Robust production of infectious hepatitis C virus (HCV) from stably HCV cDNA-transfected human hepatoma cells. J Virol 79:13963–13973

Carter RH, Fearon DT (1992) CD19: lowering the threshold for antigen receptor stimulation of B lymphocytes. Science 256:105–107

Casato M, Mecucci C, Agnello V, Fiorilli M, Knight GB, Matteucci C, Gao L, Kay J (2002) Regression of lymphoproliferative disorder after treatment for hepatitis C virus infection in a patient with partial trisomy 3, Bcl-2 overexpression, and type II cryoglobulinemia. Blood 99:2259–2261

Cerino A, Meola A, Segagni L, Furione M, Marciano S, Triyatni M, Liang TJ, Nicosia A, Mondelli MU (2001) Monoclonal antibodies with broad specificity for hepatitis C virus hypervariable region 1 variants can recognize viral particles. J Immunol 167:3878–3886

Chang KM, Rehermann B, McHutchison JG, Pasquinelli C, Southwood S, Sette A, Chisari FV (1997) Immunological significance of cytotoxic T lymphocyte epitope variants in patients chronically infected by the hepatitis C virus. J Clin Invest 100:2376–2385

Charlton M, Seaberg E, Wiesner R, Everhart J, Zetterman R, Lake J, Detre K, Hoofnagle J (1998) Predictors of patient and graft survival following liver transplantation for hepatitis C. Hepatology 28:823–830

Chaudhuri J, Tian M, Khuong C, Chua K, Pinaud E, Alt FW (2003) Transcription-targeted DNA deamination by the AID antibody diversification enzyme. Nature 422:726–730

Choisy M, Woelk CH, Guegan JF, Robertson DL (2004) Comparative study of adaptive molecular evolution in different human immunodeficiency virus groups and subtypes. J Virol 78:1962–1970

Choo QL, Kuo G, Ralston R, Weiner A, Chien D, Van Nest G, Han J, Berger K, Thudium K, Kuo C, et al (1994) Vaccination of chimpanzees against infection by the hepatitis C virus. Proc Natl Acad Sci U S A 91:1294–1298

Clayton RF, Owsianka A, Aitken J, Graham S, Bhella D, Patel AH (2002) Analysis of antigenicity and topology of E2 glycoprotein present on recombinant hepatitis C virus-like particles. J Virol 76:7672–7682

Cooper S, Erickson AL, Adams EJ, Kansopon J, Weiner AJ, Chien DY, Houghton M, Parham P, Walker CM (1999) Analysis of a successful immune response against hepatitis C virus. Immunity 10:439–449

Cooreman MP, Schoondermark-Van de Ven EM (1996) Hepatitis C virus: biological and clinical consequences of genetic heterogeneity. Scand J Gastroenterol Suppl 31:106–115

Cox AL, Mosbruger T, Mao Q, Liu Z, Wang XH, Yang HC, Sidney J, Sette A, Pardoll D, Thomas DL, Ray SC (2005) Cellular immune selection with hepatitis C virus persistence in humans. J Exp Med 201:1741–1752

Crotta S, Stilla A, Wack A, D'Andrea A, Nuti S, D'Oro U, Mosca M, Fillponi F, Brunetto RM, Bonino F, Abrignani S, Valiante NM (2002) Inhibition of natural killer cells through engagement of CD81 by the major hepatitis C virus envelope protein. J Exp Med 195:35–41

Curran R, Jameson CL, Craggs JK, Grabowska AM, Thomson BJ, Robins A, Irving WL, Ball JK (2002) Evolutionary trends of the first hypervariable region of the hepatitis C virus E2 protein in individuals with differing liver disease severity. J Gen Virol 83:11–23

Davis GL, Nelson DR, Terrault N, et al (2005) A randomized, open-label study to evaluate the safety and pharmacokinetics of human hepatitis C immune globulin (Civacir) in liver transplant recipients. Liver Transpl 11:941–949

De Francesco R, Migliaccio G (2005) Challenges and successes in developing new therapies for hepatitis C. Nature 436:953–960

De Re V, De Vita S, Marzotto A, Rupolo M, Gloghini A, Pivetta B, Gasparotto D, Carbone A, Boiocchi M (2000) Sequence analysis of the immunoglobulin antigen receptor of hepatitis C virus-associated non-Hodgkin lymphomas suggests that the malignant cells are derived from the rheumatoid factor-producing cells that occur mainly in type II cryoglobulinemia. Blood 96:3578–3584

De Vita S, De Re V, Gasparotto D, Ballare M, Pivetta B, Ferraccioli G, Pileri S, Boiocchi M, Monteverde A (2000) Oligoclonal non-neoplastic B cell expansion is the key feature of type II mixed cryoglobulinemia: clinical and molecular findings do not support a bone marrow pathologic diagnosis of indolent B cell lymphoma. Arthritis Rheum 43:94–102

Depraetere S, Leroux-Roels G (1999) Hepatitis C virus envelope proteins: immunogenicity in humans and their role in diagnosis and vaccine development. Viral Hep Rev 5:113–146

Di Noia J, Neuberger MS (2002) Altering the pathway of immunoglobulin hypermutation by inhibiting uracil-DNA glycosylase. Nature 419:43–48

Dreux M, Pietschmann T, Granier C, Voisset C, Ricard-Blum S, Mangeot PE, Keck Z, Foung S, Vu-Dac N, Dubuisson J, Bartenschlager R, Lavillette D, Cosset FL (2006) High density lipoprotein inhibits hepatitis C virus-neutralizing antibodies by stimulating cell entry via activation of the scavenger receptor BI. J Biol Chem 281:18285–18295

Drummer HE, Wilson KA, Poumbourios P (2002) Identification of the hepatitis C virus E2 glycoprotein binding site on the large extracellular loop of CD81. J Virol 76:11143–11147

Erickson AL, Kimura Y, Igarashi S, Eichelberger J, Houghton M, Sidney J, McKinney D, Sette A, Hughes AL, Walker CM (2001) The outcome of hepatitis C virus infection is predicted by escape mutations in epitopes targeted by cytotoxic T lymphocytes. Immunity 15:883–895

Everhart JE, Wei Y, Eng H, Charlton MR, Persing DH, Wiesner RH, Germer JJ, Lake JR, Zetterman RK, Hoofnagle JH (1999) Recurrent and new hepatitis C virus infection after liver transplantation. Hepatology 29:1220–1226

Faili A, Aoufouchi S, Flatter E, Gueranger Q, Reynaud CA, Weill JC (2002) Induction of somatic hypermutation in immunoglobulin genes is dependent on DNA polymerase iota. Nature 419:944–947

Farci P, Alter HJ, Govindarajan S, Wong DC, Engle R, Lesniewski RR, Mushahwar IK, Desai SM, Miller RH, Ogata N, et al (1992) Lack of protective immunity against reinfection with hepatitis C virus. Science 258:135–140

Farci P, Alter HJ, Wong DC, Miller RH, Govindarajan S, Engle R, Shapiro M, Purcell RH (1994) Prevention of hepatitis C virus infection in chimpanzees after antibody-mediated in vitro neutralization. Proc Natl Acad Sci U S A 91:7792–7796

Farci P, Shimoda A, Wong D, Cabezon T, De Gioannis D, Strazzera A, Shimizu Y, Shapiro M, Alter HJ, Purcell RH (1996) Prevention of hepatitis C virus infection in chimpanzees by hyperimmune serum against the hypervariable region 1 of the envelope 2 protein. Proc Natl Acad Sci U S A 93:15394–15399

Farci P, Shimoda A, Coiana A, Diaz G, Peddis G, Melpolder JC, Strazzera A, Chien DY, Munoz SJ, Balestrieri A, Purcell RH, Alter HJ (2000) The outcome of acute hepatitis C predicted by the evolution of the viral quasispecies. Science 288:339–344

Fearon DT, Carter RH (1995) The CD19/CR2/TAPA-1 complex of B lymphocytes: linking natural to acquired immunity. Annu Rev Immunol 13:127–149

Fitch WM, Margoliash E (1967) Construction of phylogenetic trees. Science 155:279–284

Flint M, Maidens C, Loomis-Price LD, Shotton C, Dubuisson J, Monk P, Higginbottom A, Levy S, McKeating JA (1999) Characterization of hepatitis C virus E2 glycoprotein interaction with a putative cellular receptor, CD81. J Virol 73:6235–6244

Flint M, Dubuisson J, Maidens C, Harrop R, Guile GR, Borrow P, McKeating JA (2000) Functional characterization of intracellular and secreted forms of a truncated hepatitis C virus E2 glycoprotein. J Virol 74:702–709

Forns X, Thimme R, Govindarajan S, Emerson SU, Purcell RH, Chisari FV, Bukh J (2000) Hepatitis C virus lacking the hypervariable region 1 of the second envelope protein is infectious and causes acute resolving or persistent infection in chimpanzees. Proc Natl Acad Sci USA 97:13318–13323

Foung S, Perkins S, Kafadar K, Gessner P, Zimmermann U (1990) Development of microfusion techniques to generate human hybridomas. J Immunol Methods 134:35–42

Foung SK, Perkins S, Raubitschek A, Larrick J, Lizak G, Fishwild D, Engleman EG, Grumet FC (1984) Rescue of human monoclonal antibody production from an EBV-transformed B cell line by fusion to a human-mouse hybridoma. J Immunol Methods 70:83–90

Franzin F, Efremov DG, Pozzato G, Tulissi P, Batista F, Burrone OR (1995) Clonal B-cell expansions in peripheral blood of HCV-infected patients. Br J Haematol 90:548–552

Frasca L, Del Porto P, Tuosto L, Marinari B, Scotta C, Carbonari M, Nicosia A, Piccolella E (1999) Hypervariable region 1 variants act as TCR antagonists for hepatitis C virus-specific CD4+ T cells. J Immunol 163:650–658

Fung SK, Chae HB, Fontana RJ, Conjeevaram H, Marrero J, Oberhelman K, Hussain M, Lok AS (2006) Virologic response and resistance to adefovir in patients with chronic hepatitis B. J Hepatol 44:283–290

Garcia-Retortillo M, Forns X, Feliu A, Moitinho E, Costa J, Navasa M, Rimola A, Rodes J (2002) Hepatitis C virus kinetics during and immediately after liver transplantation. Hepatology 35:680–687

Gastaminza P, Kapadia SB, Chisari FV (2006) Differential biophysical properties of infectious intracellular and secreted hepatitis C virus particles. J Virol 80:11074–11081

Germi R, Crance JM, Garin D, Guimet J, Lortat-Jacob H, Ruigrok RW, Zarski JP, Drouet E (2002) Cellular glycosaminoglycans and low density lipoprotein receptor are involved in hepatitis C virus adsorption. J Med Virol 68:206–215

Goffard A, Callens N, Bartosch B, Wychowski C, Cosset FL, Montpellier C, Dubuisson J (2005) Role of N-linked glycans in the functions of hepatitis C virus envelope glycoproteins. J Virol 79:8400–8409

Goyenechea B, Klix N, Yelamos J, Williams GT, Riddell A, Neuberger MS, Milstein C (1997) Cells strongly expressing Ig(kappa) transgenes show clonal recruitment of hypermutation: a role for both MAR and the enhancers. EMBO J 16:3987–3994

Grakoui A, Shoukry NH, Woollard DJ, Han JH, Hanson HL, Ghrayeb J, Murthy KK, Rice CM, Walker CM (2003) HCV persistence and immune evasion in the absence of memory T cell help. Science 302:659–662

Gretch DR, Polyak SJ, Wilson JJ, Carithers RL Jr, Perkins JD, Corey L (1996) Tracking hepatitis C virus quasispecies major and minor variants in symptomatic and asymptomatic liver transplant recipients. J Virol 70:7622–7631

Gruener NH, Lechner F, Jung MC, Diepolder H, Gerlach T, Lauer G, Walker B, Sullivan J, Phillips R, Pape GR, Klenerman P (2001) Sustained dysfunction of antiviral CD8+ T lymphocytes after infection with hepatitis C virus. J Virol 75:5550–5558

Guidotti LG, Chisari FV (2001) Noncytolytic control of viral infections by the innate and adaptive immune response. Annu Rev Immunol 19:65–91

Habersetzer F, Fournillier A, Dubuisson J, Rosa D, Abrignani S, Wychowski C, Nakano I, Trepo C, Desgranges C, Inchauspe G (1998) Characterization of human monoclonal antibodies specific to the hepatitis C virus glycoprotein E2 with in vitro binding neutralization properties. Virology 249:32–41

Hadlock KG, Lanford RE, Perkins S, Rowe J, Yang Q, Levy S, Pileri P, Abrignani S, Foung SK (2000) Human monoclonal antibodies that inhibit binding of hepatitis C virus E2 protein to CD81 and recognize conserved conformational epitopes. J Virol 74:10407–10416

Hanley J, Jarvis L, Simmonds P, Parker A, Ludlam C (1996) HCV and non-Hodgkin lymphoma. Lancet 347:1339

Harris SL, Craig L, Mehroke JS, Rashed M, Zwick MB, Kenar K, Toone EJ, Greenspan N, Auzanneau FI, Marino-Albernas JR, Pinto BM, Scott JK (1997) Exploring the basis of peptide-carbohydrate crossreactivity: evidence for discrimination by peptides between closely related anti-carbohydrate antibodies. Proc Natl Acad Sci U S A 94:2454–2459

Hijikata M, Shimizu YK, Kato H, Iwamoto A, Shih JW, Alter HJ, Purcell RH, Yoshikura H (1993) Equilibrium centrifugation studies of hepatitis C virus: evidence for circulating immune complexes. J Virol 67:1953–1958

Honda M, Kaneko S. Sakai A, Unoura M, Murakami S, Kobayashi K (1994) Degree of diversity of hepatitis C virus quasispecies and progression of liver disease. Hepatology 20:1144–1151

Houghton M (1996) The hepatitis C virus. In: Knight DM, Howley PM (eds) Fields virology. Lippincott–Raven, Philadelphia, pp 1035–1058

Hsu M, Zhang J, Flint M, Logvinoff C, Cheng-Mayer C, Rice CM, McKeating JA (2003) Hepatitis C virus glycoproteins mediate pH-dependent cell entry of pseudotyped retroviral particles. Proc Natl Acad Sci U S A 100:7271–7276

Ishii K, Rosa D, Watanabe Y, Katayama T, Harada H, Wyatt C, Kiyosawa K, Aizaki H, Matsuura Y, Houghton M, Abrignani S, Miyamura T (1998) High titers of antibodies inhibiting the binding of envelope to human cells correlate with natural resolution of chronic hepatitis C. Hepatology 28:1117–1120

Ivanovski M, Silvestri F, Pozzato G, Anand S, Mazzaro C, Burrone OR, Efremov DG (1998) Somatic hypermutation, clonal diversity, and preferential expression of the VH 51p1/VL kv325 immunoglobulin gene combination in hepatitis C virus-associated immunocytomas. Blood 91:2433–2442

Kaneko T, Moriyama T, Udaka K, Hiroishi K, Kita H, Okamoto H, Yagita H, Okumura K, Imawari M (1997) Impaired induction of cytotoxic T lymphocytes by antagonism of a weak agonist borne by a variant hepatitis C virus epitope. Eur J Immunol 27:1782–1787

Keck ZY, Op De Beeck A, Hadlock KG, Xia J, Li TK, Dubuisson J, Foung SK (2004a) Hepatitis C virus E2 has three immunogenic domains containing conformational epitopes with distinct properties and biological functions. J Virol 78:9224–9232

Keck ZY, Sung VM, Perkins S, Rowe J, Paul S, Liang TJ, Lai MM, Foung SK (2004b) Human monoclonal antibody to hepatitis C virus E1 glycoprotein that blocks virus attachment and viral infectivity. J Virol 78:7257–7263

Keck ZY, Li TK, Xia J, Bartosch B, Cosset FL, Dubuisson J, Foung SK (2005) Analysis of a highly flexible conformational immunogenic domain A in hepatitis C virus E2. J Virol 79: 13199–13208

Keck ZY, Xia J, Cai Z, Li TK, Owsianka AM, Patel AH, Luo G, Foung SK (2007) Immunogenic and functional organization of hepatitis C virus (HCV) glycoprotein E2 on infectious HCV virions. J Virol 81:1043–1047

Kimura Y, Gushima T, Rawale S, Kaumaya P, Walker CM (2005) Escape mutations alter proteasome processing of major histocompatibility complex class I-restricted epitopes in persistent hepatitis C virus infection. J Virol 79:4870–4876

Kizaki M, Sakashita A, Karmakar A, Lin CW, Koeffler HP (1993) Regulation of manganese superoxide dismutase and other antioxidant genes in normal and leukemic hematopoietic cells and their relationship to cytotoxicity by tumor necrosis factor. Blood 82:1142–1150

Kondo Y, Sung VM, Machida K, Liu M, Lai MM (2007) Hepatitis C virus infects T cells and affects interferon-gamma signaling in T cell lines. Virology 361:161–173

Koziel MJ, Dudley D, Afdhal N, Grakoui A, Rice CM, Choo QL, Houghton M, Walker BD (1995) HLA class I-restricted cytotoxic T lymphocytes specific for hepatitis C virus. Identification of multiple epitopes and characterization of patterns of cytokine release. J Clin Invest 96:2311–2321

Krawczynski K, Alter MJ, Tankersley DL, Beach M, Robertson BH, Lambert S, Kuo G, Spelbring JE, Meeks E, Sinha S, Carson DA (1996) Effect of immune globulin on the prevention of experimental hepatitis C virus infection. J Infect Dis 173:822–828

Kumar U, Monjardino J, Thomas HC (1994) Hypervariable region of hepatitis C virus envelope glycoprotein (E2/NS1) in an agammaglobulinemic patient. Gastroenterology 106:1072–1075

Lai ME, Mazzoleni AP, Argiolu F, De Virgilis S, Balestrieri A, Purcell RH, Cao A, Farci P (1994) Hepatitis C virus in multiple episodes of acute hepatitis in polytransfused thalassaemic children. Lancet 343:388–390

Lavillette D, Tarr AW, Voisset C, Donot P, Bartosch B, Bain C, Patel AH, Dubuisson J, Ball JK, Cosset FL (2005) Characterization of host-range and cell entry properties of the major genotypes and subtypes of hepatitis C virus. Hepatology 41:265–274

Lechner F, Wong DK, Dunbar PR, Chapman R, Chung RT, Dohrenwend P, Robbins G, Phillips R, Klenerman P, Walker BD (2000) Analysis of successful immune responses in persons infected with hepatitis C virus. J Exp Med 191:1499–1512

Levy S, Todd SC, Maecker HT (1998) CD81 (TAPA-1): a molecule involved in signal transduction and cell adhesion in the immune system. Annu Rev Immunol 16:89–109

Lindenbach BD, Rice CM (2001) Flaviviridae: the viruses and their replication. In: Knipe DM, Howley PM (eds) Fields virology. Lippincott–Raven, Philadelphia, pp 991–1041

Lindenbach BD, Evans MJ, Syder AJ, Wolk B, Tellinghuisen TL, Liu CC, Maruyama T, Hynes RO, Burton DR, McKeating JA, Rice CM (2005) Complete replication of hepatitis C virus in cell culture. Science 309:623–626

Logvinoff C, Major ME, Oldach D, Heyward S, Talal A, Balfe P, Feinstone SM, Alter H, Rice CM, McKeating JA (2004) Neutralizing antibody response during acute and chronic hepatitis C virus infection. Proc Natl Acad Sci U S A 101:10149–10154

Machida K, Cheng KT, Sung VM, Lee KJ, Levine AM, Lai MM (2004a) Hepatitis C virus infection activates the immunologic (type II) isoform of nitric oxide synthase and thereby enhances DNA damage and mutations of cellular genes. J Virol 78:8835–8843

Machida K, Cheng KT, Sung VM, Shimodaira S, Lindsay KL, Levine AM, Lai MY, Lai MM (2004b) Hepatitis C virus induces a mutator phenotype: enhanced mutations of immunoglobulin and protooncogenes. Proc Natl Acad Sci U S A 101:4262–4267

Machida K, Cheng KT, Sung VM, Levine AM, Foung S, Lai MM (2005) Hepatitis C virus E2-CD81 interaction induces hypermutation of the immunoglobulin gene in B cells. J Virol 79:8079–8089

Majid A, Jackson P, Lawal Z, Pearson GM, Parker H, Alexander GJ, Allain JP, Petrik J (1999) Ontogeny of hepatitis C virus (HCV) hypervariable region 1 (HVR1) heterogeneity and HVR1 antibody responses over a 3 year period in a patient infected with HCV type 2b. J Gen Virol 80:317–325

Major M, Rehermann B, Feinstone SM (2001) Hepatitis C viruses. In: Knipe DM, Howley PM (eds) Fields virology. Lippincott-Raven, Philadelphia, pp 1127–1162

Martell M, Esteban JI, Quer J, Vargas V, Esteban R, Guardia J, Gomez J (1994) Dynamic behavior of hepatitis C virus quasispecies in patients undergoing orthotopic liver transplantation. J Virol 68:3425–3436

McAllister J, Casino C, Davidson F, Power J, Lawlor E, Yap PL, Simmonds P, Smith DB (1998) Long-term evolution of the hypervariable region of hepatitis C virus in a common-source-infected cohort. J Virol 72:4893–4905

McKeating JA, Zhang LQ, Logvinoff C, Flint M, Zhang J, Yu J, Butera D, Ho DD, Dustin LB, Rice CM, Balfe P (2004) Diverse hepatitis C virus glycoproteins mediate viral infection in a CD81-dependent manner. J Virol 78:8496–8505

McLauchlan J, Lemberg MK, Hope G, Martoglio B (2002) Intramembrane proteolysis promotes trafficking of hepatitis C virus core protein to lipid droplets. EMBO J 21:3980–3988

Mehta SH, Cox A, Hoover DR, Wang XH, Mao Q, Ray S, Strathdee SA, Vlahov D, Thomas DL (2002) Protection against persistence of hepatitis C. Lancet 359:1478–1483

Meyer K, Basu A, Ray R (2000) Functional features of hepatitis C virus glycoproteins for pseudotype virus entry into mammalian cells. Virology 276:214–226

Meyer K, Basu A, Przysiecki CT, Lagging LM, Di Bisceglie AM, Conley AJ, Ray R (2002) Complement-mediated enhancement of antibody function for neutralization of pseudotype virus containing hepatitis C virus E2 chimeric glycoprotein. J Virol 76:2150–2158

Misiani R, Bellavita P, Fenili D, Vicari O, Marchesi D, Sironi PL, Zilio P, Vernocchi A, Massazza M, Vendramin G, et al (1994) Interferon alfa-2a therapy in cryoglobulinemia associated with hepatitis C virus. N Engl J Med 330:751–756

Miyazaki T, Muller U, Campbell KS (1997) Normal development but differentially altered proliferative responses of lymphocytes in mice lacking CD81. EMBO J 16:4217–4225

Muramatsu M, Sankaranand VS, Anant S, Sugai M, Kinoshita K, Davidson NO, Honjo T (1999) Specific expression of activation-induced cytidine deaminase (AID), a novel member of the RNA-editing deaminase family in germinal center B cells. J Biol Chem 274:18470–18476

Neumann AU, Lam NP, Dahari H, Gretch DR, Wiley TE, Layden TJ, Perelson AS (1998) Hepatitis C viral dynamics in vivo and the antiviral efficacy of interferon-alpha therapy. Science 282:103–107

Nielsen SU, Bassendine MF, Burt AD, Martin C, Pumeechockchai W, Toms GL (2006) Association between hepatitis C virus and very-low-density lipoprotein (VLDL)/LDL analyzed in iodixanol density gradients. J Virol 80:2418–2428

Op De Beeck A, Voisset C, Bartosch B, Ciczora Y, Cocquerel L, Keck Z, Foung S, Cosset FL, Dubuisson J (2004) Characterization of functional hepatitis C virus envelope glycoproteins. J Virol 78:2994–3002

Owsianka A, Clayton RF, Loomis-Price LD, McKeating JA, Patel AH (2001) Functional analysis of hepatitis C virus E2 glycoproteins and virus-like particles reveals structural dissimilarities between different forms of E2. J Gen Virol 82:1877–1883

Owsianka A, Tarr AW, Juttla VS, Lavillette D, Bartosch B, Cosset FL, Ball JK, Patel AH (2005) Monoclonal antibody AP33 defines a broadly neutralizing epitope on the hepatitis C virus E2 envelope glycoprotein. J Virol 79:11095–11104

Owsianka AM, Timms JM, Tarr AW, Brown RJ, Hickling TP, Szwejk A, Bienkowska-Szewczyk K, Thomson BJ, Patel AH, Ball JK (2006) Identification of conserved residues in the E2 envelope glycoprotein of the hepatitis C virus that are critical for CD81 binding. J Virol 80:8695–8704

Papavasiliou FN, Schatz DG (2002) Somatic hypermutation of immunoglobulin genes: merging mechanisms for genetic diversity. Cell 109 Suppl:S35–44

Penin F, Combet C, Germanidis G, Frainais PO, Deleage G, Pawlotsky JM (2001) Conservation of the conformation and positive charges of hepatitis C virus E2 envelope glycoprotein hypervariable region 1 points to a role in cell attachment. J Virol 75:5703–5710

Perkins S, Zimmermann U, Foung SK (1991) Parameters to enhance human hybridoma formation with hypoosmolar electrofusion. Hum Antibodies Hybridomas 2:155–159

Perrillo RP, Wright T, Rakela J, et al (2001) A multicenter United States-Canadian trial to assess lamivudine monotherapy before and after liver transplantation for chronic hepatitis B. Hepatology 33:424–432

Pestka JM, Zeisel MB, Bläser E, Schürmann P, Bartosch B, Cosset F-L, Patel AH, Meisel H, Baumert J, Viazov S, Rispeter K, Blum HE, Roggendorf M, Baumert TF (2007) Rapid induction of virus-neutralizing antibodies and viral clearance in a single-source outbreak of hepatitis C. Proc Natl Acad Sci U S A 104:6025–6030

Petersen-Mahrt SK, Harris RS, Neuberger MS (2002) AID mutates E. coli suggesting a DNA deamination mechanism for antibody diversification. Nature 418:99–103

Piazza M, Sagliocca L, Tosone G, Guadagnino V, Stazi MA, Orlando R, Borgia G, Rosa D, Abrignani S, Palumbo F, Manzin A, Clementi M (1997) Sexual transmission of the hepatitis C virus and efficacy of prophylaxis with intramuscular immune serum globulin. A randomized controlled trial. Arch Intern Med 157:1537–1544

Pietschmann T, Kaul A, Koutsoudakis G, Shavinskaya A, Kallis S, Steinmann E, Abid K, Negro F, Dreux M, Cosset FL, Bartenschlager R (2006) Construction and characterization of infectious intragenotypic and intergenotypic hepatitis C virus chimeras. Proc Natl Acad Sci USA 103:7408–7413

Pileri P, Uematsu Y, Campagnoli S, Galli G, Falugi F, Petracca R, Weiner AJ, Houghton M, Rosa D, Grandi G, Abrignani S (1998) Binding of hepatitis C virus to CD81. Science 282:938–941

Pinter A, Honnen WJ, He Y, Gorny MK, Zolla-Pazner S, Kayman SC (2004) The V1/V2 domain of gp120 is a global regulator of the sensitivity of primary human immunodeficiency virus type

1 isolates to neutralization by antibodies commonly induced upon Infection. J Virol 78:5205–5215

Post JJ, Pan Y, Freeman AJ, et al (2004) Clearance of hepatitis C viremia associated with cellular immunity in the absence of seroconversion in the hepatitis C incidence and transmission in prisons study cohort. J Infect Dis 189:1846–1855

Pugach P, Kuhmann SE, Taylor J, Marozsan AJ, Snyder A, Ketas T, Wolinsky SM, Korber BT, Moore JP (2004) The prolonged culture of human immunodeficiency virus type 1 in primary lymphocytes increases its sensitivity to neutralization by soluble CD4. Virology 321:8–22

Puig M, Major ME, Mihalik K, Feinstone SM (2004) Immunization of chimpanzees with an envelope protein-based vaccine enhances specific humoral and cellular immune responses that delay hepatitis C virus infection. Vaccine 22:991–1000

Puntoriero G, Meola A, Lahm A, Zucchelli S, Ercole BB, Tafi R, Pezzanera M, Mondelli MU, Cortese R, Tramontano A, Galfre' G, Nicosia A (1998) Towards a solution for hepatitis C virus hypervariability: mimotopes of the hypervariable region 1 can induce antibodies cross-reacting with a large number of viral variants. EMBO J 17:3521–3533

Rada C, Milstein C (2001) The intrinsic hypermutability of antibody heavy and light chain genes decays exponentially. EMBO J 20:4570–4576

Radkowski M, Bednarska A, Horban A, Stanczak J, Wilkinson J, Adair DM, Nowicki M, Rakela J, Laskus T (2004) Infection of primary human macrophages with hepatitis C virus in vitro: induction of tumour necrosis factor-alpha and interleukin 8. J Gen Virol 85:47–59

Ray SC, Wang YM, Laeyendecker O, Ticehurst JR, Villano SA, Thomas DL (1999) Acute hepatitis C virus structural gene sequences as predictors of persistent viremia: hypervariable region 1 as a decoy. J Virol 73:2938–2946

Ray SC, Fanning L, Wang XH, Netski DM, Kenny-Walsh E, Thomas DL (2005) Divergent and convergent evolution after a common-source outbreak of hepatitis C virus. J Exp Med 201:1753–1759

Robertson B, Myers G, Howard C, et al (1998) Classification, nomenclature, and database development for hepatitis C virus (HCV) and related viruses: proposals for standardization. International Committee on Virus Taxonomy. Arch Virol 143:2493–2503

Roccasecca R, Folgori A, Ercole BB, Puntoriero G, Lahm A, Zucchelli S, Tafi R, Pezzanera M, Galfre G, Tramontano A, Mondelli MU, Pessi A, Nicosia A, Cortese R, Meola A (2001) Mimotopes of the hyper variable region 1 of the hepatitis C virus induce cross-reactive antibodies directed against discontinuous epitopes. Mol Immunol 38:485–492

Rollier C, Depla E, Drexhage JA, Verschoor EJ, Verstrepen BE, Fatmi A, Brinster C, Fournillier A, Whelan JA, Whelan M, Jacobs D, Maertens G, Inchauspe G, Heeney JL (2004) Control of heterologous hepatitis C virus infection in chimpanzees is associated with the quality of vaccine-induced peripheral T-helper immune response. J Virol 78:187–196

Rosa D, Campagnoli S, Moretto C, Guenzi E, Cousens L, Chin M, Dong C, Weiner AJ, Lau JY, Choo QL, Chien D, Pileri P, Houghton M, Abrignani S (1996) A quantitative test to estimate neutralizing antibodies to the hepatitis C virus: cytofluorimetric assessment of envelope glycoprotein 2 binding to target cells. Proc Natl Acad Sci U S A 93:1759–1763

Rushbrook SM, Ward SM, Unitt E, Vowler SL, Lucas M, Klenerman P, Alexander GJ (2005) Regulatory T cells suppress in vitro proliferation of virus-specific CD8[+] T cells during persistent hepatitis C virus infection. J Virol 79:7852–7859

Sanchez-Quijano A, Pineda JA, Lissen E, Leal M, Diaz-Torres MA, Garcia De Pesquera F, Rivera F, Castro R, Munoz J (1988) Prevention of post-transfusion non-A, non-B hepatitis by non-specific immunoglobulin in heart surgery patients. Lancet 1:1245–1249

Sasso EH (2000) The rheumatoid factor response in the etiology of mixed cryoglobulins associated with hepatitis C virus infection. Ann Med Interne (Paris) 151:30–40

Scarselli E, Ansuini H, Cerino R, Roccasecca RM, Acali S, Filocamo G, Traboni C, Nicosia A, Cortese R, Vitelli A (2002) The human scavenger receptor class B type I is a novel candidate receptor for the hepatitis C virus. EMBO J 21:5017–5025

Schofield DJ, Bartosch B, Shimizu YK, Allander T, Alter HJ, Emerson SU, Cosset FL, Purcell RH (2005) Human monoclonal antibodies that react with the E2 glycoprotein of hepatitis C virus and possess neutralizing activity. Hepatology 42:1055–1062

Schreibman IR, Schiff ER (2006) Prevention and treatment of recurrent hepatitis B after liver transplantation: the current role of nucleoside and nucleotide analogues. Ann Clin Microbiol Antimicrob 5:8

Seifert U, Liermann H, Racanelli V, Halenius A, Wiese M, Wedemeyer H, Ruppert T, Rispeter K, Henklein P, Sijts A, Hengel H, Kloetzel PM, Rehermann B (2004) Hepatitis C virus mutation affects proteasomal epitope processing. J Clin Invest 114:250–259

Sheridan I, Pybus OG, Holmes EC, Klenerman P (2004) High-resolution phylogenetic analysis of hepatitis C virus adaptation and its relationship to disease progression. J Virol 78:3447–3454

Shimizu YK, Hijikata M, Iwamoto A, Alter HJ, Purcell RH, Yoshikura H (1994) Neutralizing antibodies against hepatitis C virus and the emergence of neutralization escape mutant viruses. J Virol 68:1494–1500

Shimizu YK, Igarashi H, Kiyohara T, Cabezon T, Farci P, Purcell RH, Yoshikura H (1996) A hyperimmune serum against a synthetic peptide corresponding to the hypervariable region 1 of hepatitis C virus can prevent viral infection in cell cultures. Virology 223:409–412

Shoukry NH, Cawthon AG, Walker CM (2004) Cell-mediated immunity and the outcome of hepatitis C virus infection. Annu Rev Microbiol 58:391–424

Siemoneit K, Cardoso Mda S, Koerner K, Wolpl A, Kubanek B (1995) Human monoclonal antibodies for the immunological characterization of a highly conserved protein domain of the hepatitis C virus glycoprotein E1. Clin Exp Immunol 101:278–283

Silvestri F, Barillari G, Fanin R, Pipan C, Falasca E, Salmaso F, Zaja F, Infanti L, Patriarca F, Botta GA, Baccarani M (1997) The genotype of the hepatitis C virus in patients with HCV-related B cell non-Hodgkin's lymphoma. Leukemia 11:2157–2161

Simmonds P, Bukh J, Combet C, et al (2005) Consensus proposals for a unified system of nomenclature of hepatitis C virus genotypes. Hepatology 42:962–973

Smith DB (1999) Evolution of the hypervariable region of hepatitis C virus. J Viral Hepat 6 [Suppl 1]:41–46

Smith TJ (2001) Antibody interactions with rhinovirus: lessons for mechanisms of neutralization and the role of immunity in viral evolution. Curr Top Microbiol Immunol 260:1–28

Sneath PHA, Sokal RR (1973) Taxonomic structure. In: Numerical taxonomy. Freeman and Company, San Francisco, pp 230–234

Soldaini E, Wack A, D'Oro U, Nuti S, Ulivieri C, Baldari CT, Abrignani S (2003) T cell costimulation by the hepatitis C virus envelope protein E2 binding to CD81 is mediated by Lck. Eur J Immunol 33:455–464

Steinmann D, Barth H, Gissler B, Schurmann P, Adah MI, Gerlach JT, Pape GR, Depla E, Jacobs D, Maertens G, Patel AH, Inchauspe G, Liang TJ, Blum HE, Baumert TF (2004) Inhibition of hepatitis C virus-like particle binding to target cells by antiviral antibodies in acute and chronic hepatitis C. J Virol 78:9030–9040

Sugg U, Schneider W, Hoffmeister HE, Huth C, Stephan W, Lissner R, Haase W (1985) Hepatitis B immune globulin to prevent non-A, non-B post-transfusion hepatitis. Lancet 1:405–406

Sullivan J, Swofford DL, Naylor JP (1999) The effect of taxon sampling on estimating rate heterogeneity parameters of maximum-likelihood models. Mol Biol Evol 16:1347–1356

Sung VM, Shimodaira S, Doughty AL, Picchio GR, Can H, Yen TS, Lindsay KL, Levine AM, Lai MM (2003) Establishment of B-cell lymphoma cell lines persistently infected with hepatitis C virus in vivo and in vitro: the apoptotic effects of virus infection. J Virol 77:2134–2146

Takahashi S, Doss C, Levy S, Levy R (1990) TAPA-1, the target of an antiproliferative antibody, is associated on the cell surface with the Leu-13 antigen. J Immunol 145:2207–2213

Tarr AW, Owsianka AM, Timms JM, McClure CP, Brown RJ, Hickling TP, Pietschmann T, Bartenschlager R, Patel AH, Ball JK (2006) Characterization of the hepatitis C virus E2 epitope defined by the broadly neutralizing monoclonal antibody AP33. Hepatology 43:592–601

Tester I, Smyk-Pearson S, Wang P, Wertheimer A, Yao E, Lewinsohn DM, Tavis JE, Rosen HR (2005) Immune evasion versus recovery after acute hepatitis C virus infection from a shared source. J Exp Med 201:1725–1731

Timm J, Lauer GM, Kavanagh DG, Sheridan I, Kim AY, Lucas M, Pillay T, Ouchi K, Reyor LL, Schulze zur Wiesch J, Gandhi RT, Chung RT, Bhardwaj N, Klenerman P, Walker BD, Allen TM (2004) CD8 epitope escape and reversion in acute HCV infection. J Exp Med 200:1593–1604

Triyatni M, Vergalla J, Davis AR, Hadlock KG, Foung SK, Liang TJ (2002) Structural features of envelope proteins on hepatitis C virus-like particles as determined by anti-envelope monoclonal antibodies and CD81 binding. Virology 298:124–132

Tsai SL, Chen YM, Chen MH, Huang CY, Sheen IS, Yeh CT, Huang JH, Kuo GC, Liaw YF (1998) Hepatitis C virus variants circumventing cytotoxic T lymphocyte activity as a mechanism of chronicity. Gastroenterology 115:954–965

Tseng CT, Klimpel GR (2002) Binding of the hepatitis C virus envelope protein E2 to CD81 inhibits natural killer cell functions. J Exp Med 195:43–49

Voisset C, Op de Beeck A, Horellou P, Dreux M, Gustot T, Duverlie G, Cosset FL, Vu-Dac N, Dubuisson J (2006) High-density lipoproteins reduce the neutralizing effect of hepatitis C virus (HCV)-infected patient antibodies by promoting HCV entry. J Gen Virol 87:2577–2581

von Hahn T, Yoon JC, Alter H, Rice CM, Rehermann B, Balfe P, McKeating JA (2007) Hepatitis C virus continuously escapes from neutralizing antibody and T-cell responses during chronic infection in vivo. Gastroenterology 132:667–678

Wakita T, Pietschmann T, Kato T, Date T, Miyamoto M, Zhao Z, Murthy K, Habermann A, Krausslich HG, Mizokami M, Bartenschlager R, Liang TJ (2005) Production of infectious hepatitis C virus in tissue culture from a cloned viral genome. Nat Med 11:791–796

Wedemeyer H, He XS, Nascimbeni M, Davis AR, Greenberg HB, Hoofnagle JH, Liang TJ, Alter H, Rehermann B (2002) Impaired effector function of hepatitis C virus-specific CD8+ T cells in chronic hepatitis C virus infection. J Immunol 169:3447–3458

Weiner A, Erickson AL, Kansopon J, Crawford K, Muchmore E, Hughes AL, Houghton M, Walker CM (1995) Persistent hepatitis C virus infection in a chimpanzee is associated with emergence of a cytotoxic T lymphocyte escape variant. Proc Natl Acad Sci USA 92:2755–2759

Weng WK, Levy S (2003) Hepatitis C virus (HCV) and lymphomagenesis. Leuk Lymphoma 44:1113–1120

Wong GH, Goeddel DV (1988) Induction of manganous superoxide dismutase by tumor necrosis factor: possible protective mechanism. Science 242:941–944

Yang Z, Bielawski JP (2000) Statistical methods for detecting molecular adaptation. Trends Ecol Evol 15:496–503

Yi M, Villanueva RA, Thomas DL, Wakita T, Lemon SM (2006) Production of infectious genotype 1a hepatitis C virus (Hutchinson strain) in cultured human hepatoma cells. Proc Natl Acad Sci U S A 103:2310–2315

Zan H, Komori A, Li Z, Cerutti A, Schaffer A, Flajnik MF, Diaz M, Casali P (2001) The translesion DNA polymerase zeta plays a major role in Ig and bcl-6 somatic hypermutation. Immunity 14:643–653

Zeng X, Winter DB, Kasmer C, Kraemer KH, Lehmann AR, Gearhart PJ (2001) DNA polymerase eta is an A-T mutator in somatic hypermutation of immunoglobulin variable genes. Nat Immunol 2:537–541

Zhang J, Randall G, Higginbottom A, Monk P, Rice CM, McKeating JA (2004) CD81 is required for hepatitis C virus glycoprotein-mediated viral infection. J Virol 78:1448–1455

Zhong J, Gastaminza P, Cheng G, Kapadia S, Kato T, Burton DR, Wieland SF, Uprichard SL, Wakita T, Chisari FV (2005) Robust hepatitis C virus infection in vitro. Proc Natl Acad Sci U S A 102:9294–9299

Zhou YH, Shimizu YK, Esumi M (2000) Monoclonal antibodies to the hypervariable region 1 of hepatitis C virus capture virus and inhibit virus adsorption to susceptible cells in vitro. Virology 269:276–283

Zibert A, Schreier E, Roggendorf M (1995) Antibodies in human sera specific to hypervariable region 1 of hepatitis C virus can block viral attachment. Virology 208:653–661

Zibert A, Kraas W, Meisel H, Jung G, Roggendorf M (1997) Epitope mapping of antibodies directed against hypervariable region 1 in acute self-limiting and chronic infections due to hepatitis C virus. J Virol 71:4123–4127

Zimmermann U, Gessner P, Schnettler R, Perkins S, Foung SK (1990) Efficient hybridization of mouse-human cell lines by means of hypo-osmolar electrofusion. J Immunol Methods 134:43–50

Zucchelli S, Roccasecca R, Meola A, Ercole BB, Tafi R, Dubuisson J, Galfre G, Cortese R, Nicosia A (2001) Mimotopes of the hepatitis C virus hypervariable region 1, but not the natural sequences, induce cross-reactive antibody response by genetic immunization. Hepatology 33:692–703

Zuckerman E, Zuckerman T, Levine AM, Douer D, Gutekunst K, Mizokami M, Qian DG, Velankar M, Nathwani BN, Fong TL (1997) Hepatitis C virus infection in patients with B-cell non-Hodgkin lymphoma. Ann Intern Med 127:423–428

Antibodies for HIV Treatment and Prevention: Window of Opportunity?

M. Huber, W.C. Olson, and A. Trkola(✉)

Abstract Monoclonal antibodies are routinely used as therapeutics in a number of disease settings and have thus also been explored as potential treatment for human immunodeficiency virus (HIV)-1 infection. Antibodies targeting viral antigens, and those directed to the cellular receptors, have been considered for use in prevention and therapy. For virus-targeted antibodies, attention has focused primarily on their neutralizing activity, but such antibodies also have the potential to exert antiviral effects via effector functions, such as antibody-dependent cellular cytotoxicity (ADCC), opsonization, or complement activation. Anti-cell antibodies act through occlusion or down-modulation of the viral receptors with notable impact in vivo, as recent trials have shown. This review summarizes the diverse specificities and modes of action of therapeutic antibodies against HIV-1 infection. Successes, challenges, and future opportunities of harnessing antibodies for therapy of HIV-1 infection are discussed.

A. Trkola
Division of Infectious Diseases, University Hospital Zurich, Rämistrasse 100, 8091 Zurich, Switzerland
e-mail: alexandra.trkola@usz.ch

S.K. Dessain (ed.) *Human Antibody Therapeutics for Viral Disease. Current Topics in Microbiology and Immunology 317.*
© Springer-Verlag Berlin Heidelberg 2008

1   Introduction

Twenty-five years into the human immunodeficiency virus (HIV) epidemic, antiretroviral therapy (ART), where available, succeeds in dramatically reducing mortality and morbidity and significantly lowers rates of mother-to-child transmission (MTCT). However, although initiation of ART suppresses viral loads to undetectable levels for several years in most individuals (Gulick et al. 2006), hopes that HIV infection would eventually be cleared by ART have not yet been fulfilled. Consequently, infected individuals may require life-long treatment, which can be problematic due to side effects of the drugs and evolution of viral resistance. Emergence of drug-resistant strains has been reported for all currently licensed substances and necessitates the unremitting development of alternate therapies. While oral drugs are in general the preferred choice, due to their ease of administration, antibodies that interfere with viral replication have also been considered for therapeutic purposes in HIV infection.

To date, polyclonal and monoclonal antibody therapeutics are routinely used in cancer therapy and diagnostics, in autoimmune disorders, as antitoxins, and in the treatment or prevention of viral, bacterial, or parasitic infections (Keller et al. 2000; Sawyer 2000; Zeitlin et al. 2000; Brekke et al. 2003; Reichert et al. 2005; Schrama et al. 2006). Over 20 monoclonal antibodies (mAbs) and immunoglobulin Fc fusion proteins have received FDA approval. In a number of instances these products represent first-line therapy and the current standard of care. In viral infections, polyclonal and monoclonal antibodies are used for treatment and prevention of infections with hepatitis B virus (HBV), cytomegalovirus, varicella zoster virus, respiratory syncytial virus, and rabies virus (Sawyer 2000; Brekke et al. 2003). For example, Synagis (palivizumab; MedImmune, Gaithersburg, MD), a humanized IgG1 mAb to the respiratory syncytial virus (RSV) fusion protein, remains the product of choice for prevention of serious lower respiratory tract disease caused by RSV in children, despite intensive research to develop small molecule inhibitors (Ding et al. 1998; McKimm-Breschkin 2000; Huntley et al. 2002; Cianci et al. 2004).

The potential that antibody-based therapeutics bear in treatment and prevention of HIV infection has been zealously debated over the years. Do we have enough evidence that naturally occurring antibodies impact on HIV infection to support this approach? Have clinical studies of investigational antibodies provided clear proof-of-concept? Which antibodies would we need to develop? In which clinical settings could antibody-based therapeutics be of greatest use? In principle, a wide spectrum of antibodies with diverse specificities and modes of action could be envisioned for therapeutic purposes in HIV infection (see Fig. 1 and Table 1). Antibodies that are directed against both the virus and cellular receptors have demonstrated activity against HIV and can block its infectivity. In the following we will summarize the knowledge gained on the functionality and feasibility of antibody therapeutics in HIV infection over the years and emphasize areas that await further investigation.

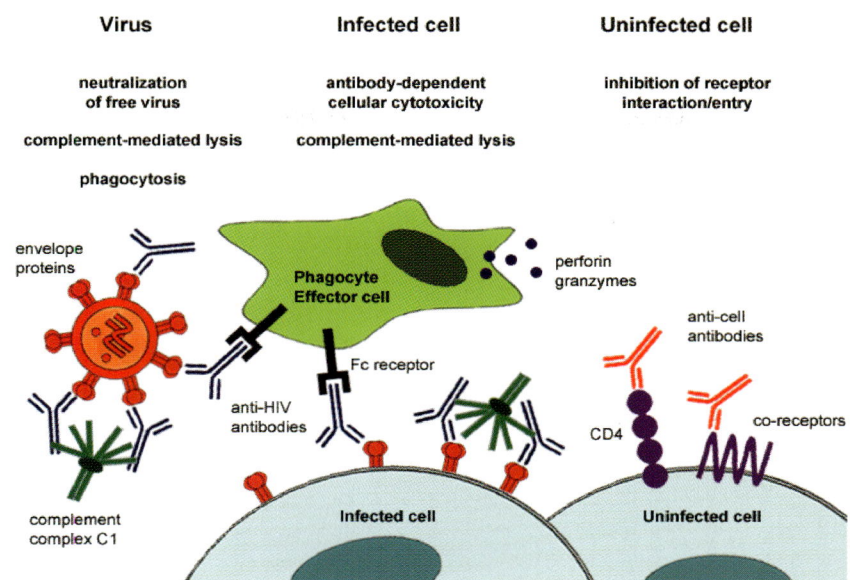

Fig. 1 Modes of action of therapeutic antibodies in prevention and therapy. Anti-HIV antibodies (*blue*), anti-cell antibodies (*red*), CD4 and co-receptors (*violet*), Fc receptors (*black*)

2 Targeting the Virus and Infected Cells

HIV infects host cells upon interaction of the viral envelope glycoprotein gp120 with the cellular receptor CD4 (Maddon et al. 1986) and a co-receptor (most commonly CCR5 or CXCR4) (Alkhatib et al. 1996; Deng et al. 1996; Dragic et al. 1996; Feng et al. 1996). Receptor binding induces conformational changes in gp120 that subsequently lead to rearrangements in gp41, the transmembrane unit of the envelope, and prompt fusion (Wyatt et al. 1998b; Pierson et al. 2003).

Antibodies employ distinct modes of action to interfere with the life cycle of viruses. The initial steps in viral infection—receptor engagement and fusion mediated by the envelope glycoproteins—are primary targets for neutralizing antibodies (Klasse et al. 2002). All neutralizing antibodies against HIV described to date inhibit infection of target cells by blocking engagement of CD4 or the co-receptor, or by binding to domains involved in subsequent steps of the fusion process (Trkola et al. 1996a; Wu et al. 1996; Wyatt et al. 1998b; Parren et al. 2001a; Xiang et al. 2002; Decker et al. 2005; Pantophlet et al. 2006). The initially high hopes of exploiting antibody-based immunity for treatment and prevention were dampened when it became clear that most of the neutralizing activity elicited to HIV-1 in vivo is strain and subtype specific. Demands on therapeutic antibodies are severe: they have to be safe, potent, and broadly active against divergent HIV strains. To date, only a handful of monoclonal antibodies that neutralize with a

Table 1 Potential functions of antibody therapeutics in HIV infection

Target	Virus	Infected cell	Uninfected cell
Epitope	HIV envelope proteins (gp120, gp41)	HIV envelope proteins (gp120, gp41)	HIV receptors (CD4, CCR5, CXCR4)
	Virus subtype specific	Virus subtype specific	No subtype specificity
Mechanisms of action	Neutralization (inhibition of entry and fusion)	Antibody-dependent cellular cytotoxicity (ADCC)	Inhibition of receptor interaction/entry
	Fc and CR-mediated phagocytosis	Complement-mediated lysis	
	Complement-mediated lysis		
Effector functions	Potentially advantageous	Potentially advantageous	Potentially detrimental
Antibody class	IgG (all subtypes), IgM, IgA	IgG (all subtypes), IgM, IgA	IgG2 and IgG4 preferred to limit effector functions
Potential use in prevention	Microbicide		Microbicide
	Mother-to-child transmission		Mother-to-child transmission
	Active/passive immunization		Post-exposure prophylaxis
Potential use in therapy	Combination therapy as part of ART regimen	Combination therapy as part of ART regimen	Combination therapy as part of ART regimen
	Intermittent treatment during drug holiday	Combination therapy with early ART to reduce pool of infected cells	Intermittent treatment during drug holiday
	Diagnostics	Immunotoxin (combined with other strategies to reactivate/eliminate latent reservoir)	
		Diagnostics	
Escape mechanisms	Rapid escape through mutations in viral proteins	Rapid escape through mutations in viral proteins	Escape (virus changes binding site on receptor)
Potential safety concerns	Antibody-dependent enhancement of infection		Interference with immune functions and cell depletion
Tested in humans	2G12[a]	Effector function of anti-HIV antibodies have not been verified in vivo	TNX-355[e]
	2F5[b]		PRO 140[f]
	4E10[c]		CCR5mAb004[g]
	F105[d]		

CR, complement receptor

[a]Armbruster et al. 2002; Stiegler et al. 2002; Trkola et al. 2005; [b]Armbruster et al. 2002, 2004; Stiegler et al. 2002; Trkola et al. 2005; [c]Armbruster et al. 2002, 2004; Trkola et al. 2005; [d]Wolfe et al. 1996; Cavacini et al. 1998; [e]Jacobson et al. 2004b; Norris et al. 2006; [f]Olson et al. 2006; [g]Roschke et al. 2004

broad cross-neutralizing activity have been isolated: the antibody IgG1b12, which recognizes a unique epitope overlapping the CD4-binding site (Burton et al. 1994), the carbohydrate-specific antibody 2G12 (Trkola et al. 1996b), which recognizes an equally unique mannose-dependent epitope within gp120 (Scanlan et al. 2003), and the antibodies 2F5 and 4E10, which bind to the membrane-proximal external region (MPER) in gp41 (Muster et al. 1993; Stiegler et al. 2001; Zwick et al. 2001; Binley et al. 2004).

2.1   Modes of Action of Antiviral Antibodies in HIV Infection

Initial efforts to develop antibodies for therapeutic use focused on defining antibodies with neutralizing capacity—i.e., antibodies that bind virus and inhibit entry into target cells. As discussed above, neutralizing antibodies and their mechanisms of action have been intensively studied over the years. It is generally agreed that antibodies with neutralizing capacity will be important components of vaccine-induced immunity and thus have also been prime candidates for the development as therapeutics. All HIV-specific antibodies probed for efficacy in vivo to date have neutralizing activities. Their characteristics and effects will be discussed in detail below.

Whether neutralizing activity is the sole function these antibodies can or should fulfill in vivo has been increasingly debated. Besides neutralizing free viruses, antibodies could have significant impact on virus elimination by inducing phagocytosis or complement-dependent lysis of opsonized viral particles (Fig. 1). Activity of complement and antibody in controlling viral infection has been described in other viral diseases (Pincus et al. 1995; Blue et al. 2004; Hangartner et al. 2006). However, to what extent this mechanism is active against HIV remains uncertain: antibodies against HIV that induce complement lysis of virions are common in HIV infection (Aasa-Chapman et al. 2005; Huber et al. 2006) and may contribute to viral control in vivo (Huber et al. 2006). Nevertheless, the overall lysis activity against HIV is low compared to other viruses (Marschang et al. 1993; Sullivan et al. 1996; Stoiber et al. 2001; Huber et al. 2006). HIV weakens complement attack by incorporating into its outer membrane the cellular complement regulatory proteins CD46, CD55, and CD59, which cause termination of the complement cascade and rescue the virus from lysis (Montefiori et al. 1994; Marschang et al. 1995; Saifuddin et al. 1997). Furthermore, antibody- and complement-coated virions could potentially enhance infection of Fc and complement receptor-expressing cells, which needs to be considered if antibodies that trigger effector functions are used for therapy (Robinson et al. 1988; Takeda et al. 1988; Montefiori 1997; Stoiber et al. 2001; Stoiber et al. 2005). Of note, none of the passive immunization studies conducted to date has given evidence of antibody-driven enhancement in vivo. That is, treatment with antibody resulted in decreased or unchanged viral loads; increased viral burden was not observed (Gunthard et al. 1994; Mascola et al. 2000; Armbruster et al. 2004; Trkola et al. 2005). Nevertheless, whether or not vaccines or therapeutics should include antibodies that elicit complement activity needs to be carefully evaluated to assess potential benefits and risks of their activity.

Another major immune function of antiviral antibodies that needs to be evaluated is antibody-dependent cellular cytotoxicity (ADCC). If active in HIV infection, the impact of this mechanism could indeed be high, as ADCC-mediating antibodies would eliminate HIV-1 infected cells and thereby reduce production of progeny. Antibodies that mediate ADCC are omnipresent in HIV-infected individuals, although uncertainty prevails on the functionality of the mechanism, as effector cells (natural killer cells, neutrophils) may have decreased activity upon disease progression (Bender et al. 1988; Monari et al. 1999; Azzam et al. 2006). Evidence has accumulated that ADCC-mediating antibodies may nevertheless function in vivo, as they were described to positively influence disease progression (Baum et al. 1996; Broliden et al. 1993; Ahmad et al. 2001; Forthal et al. 2001). Equally, in animal models ADCC activity has been associated with delayed disease progression and protection (Belo et al. 1991; Ferrari et al. 1994; Broliden et al. 1996; Banks et al. 2002). Antibody therapeutics that function through complement and/or ADCC are used in cancer treatment (Golay et al. 2003; Mimura et al. 2005; van Meerten et al. 2006), but have not yet been actively pursued for treatment in HIV infection. This may need reconsideration, as the effect on HIV replication could be significant if ADCC could be successfully harnessed to destroy infected cells. Latent reservoirs of HIV-infected cells with extremely long half-lives have been identified (Finzi et al. 1997, 1999; Wong et al. 1997), a situation excluding the possibility that HIV infection in a patient can be eradicated within a frame of several years, as initially proposed (Perelson et al. 1997). Attempts to eliminate this latent reservoir in vivo through stimulation paired with ART have failed so far (Kulkosky et al. 2006). ADCC-mediating antibodies could be envisioned to be effective in such combined stimulation/elimination strategies, supporting their further investigation.

Antibodies that specifically target virus and infected cells could be further used for generation of immunotoxins, an approach that has proved successful in cancer treatment (Bross et al. 2001; Pastan et al. 2006; Schrama et al. 2006). Immunotoxins are antibodies linked to toxins that, when taken up by target cells, lead to their destruction. Several immunotoxins that specifically eliminate HIV-infected cells in vitro have been developed in previous years (Till et al. 1989; Pincus et al. 1990, 1991, 1993; Pincus 1996; Lueders et al. 2004). Clinical testing of the immunotoxin CD4(178)PE40, a fusion protein directed against the CD4 binding site of gp120, showed little effect, however, which was attributed to rapid clearance of the immunotoxin and the differential resistance of clinical HIV isolates (Davey et al. 1994; Ramachandran et al. 1994). Recent studies suggest that combination of antibodies of different specificities and other toxins could potentially improve efficacy and tolerability of immunotoxins (Johansson et al. 2006; Kennedy et al. 2006).

Obviously, as with all medications, safety concerns are high for antibody therapeutics. As mentioned already, antibody and complement may lead to infection enhancement. Eliciting ADCC bears the potential of harming uninfected cells if antibodies are polyspecific or viral antigen bound to uninfected cells is recognized. The latter has been shown in vitro, where uninfected cells coated with gp120 were susceptible to ADCC (Lyerly et al. 1987; Ahmad et al. 1994; Hober et al. 1995). While free gp120 in vivo is not likely to be present at high enough concentrations for this to occur (Klasse et al. 2004), this scenario needs to be considered. Polyspecificity and autoreactivity of antibodies, and the potentially ensuing adverse effects, came into recent focus when it was described that 2F5, 4E10, and to a lesser extent IgG1b12 cross react with autoantigens (Haynes et al. 2005), prompting concern that these antibodies upon in vivo application may predispose patients for autoimmunity. Notably though, in the case of 2F5 and 4E10, prolonged treatment at high doses during phase I and II testing in adults showed no serious adverse effects (Trkola et al. 2005). Moreover, there was no sign of autoimmune disease despite high and sustained levels of these two antibodies. Nevertheless, considering the in vitro data available, monitoring of patients for autoimmune disease would be appropriate in future investigational trials with these antibodies.

While the main focus has been on defining neutralizing antibodies that limit transmission of free virus particles, the role free virions play in vivo compared to spread from cell to cell remains ambiguous. Notably, in vitro the capacity of neutralizing antibodies in limiting cell-to-cell spread was reported to be considerably lower than the activity against free virions (Pantaleo et al. 1995). Only recently it became evident that in close cellular contact HIV particles spread from cell to cell via so-called virological synapses (Sato et al. 1992; Bangham 2003; McDonald et al. 2003; Arrighi et al. 2004; Jolly et al. 2004a, b). Transmission through the synapse may render the virus less susceptible to neutralization; however, no change in susceptibility was observed for virus transmitted *in trans* across the synapse between dendritic cells and T cells (Ketas et al. 2003). Over the coming years, it will be important to determine the modes of viral transmission in vivo and whether these modes of transmission can be effectively blocked by neutralizing antibodies. This information may be important for designing effective vaccines and antibody therapeutics.

2.2 Efficacy of HIV-Specific Antibodies in vivo

Passive immunization studies in animal models conducted over recent years brought the confirmation that neutralizing antibodies function in vivo and can limit transmission and de novo infection when applied topically (Veazey et al. 2003) or systemically (Gauduin et al. 1997; Shibata et al. 1999; Baba et al. 2000; Mascola et al. 2000; Hofmann-Lehmann et al. 2001; Montefiori et al. 2001; Parren et al. 2001b; Ruprecht et al. 2001; Haigwood et al. 2004).

Among the broadly neutralizing HIV antibodies, so far only 2G12, 2F5, and 4E10 have undergone clinical testing and will thus be discussed below in more detail. The discussion also includes PRO 542 (previously referred to as CD4-IgG2), a tetravalent CD4-immunoglobulin fusion protein that also broadly neutralizes HIV (Allaway et al. 1995). The first monoclonal antibody against HIV applied in passive immunization in vivo was MAb F105, which binds to the CD4-binding site (Posner et al. 1991). However this antibody has a comparatively restricted neutralization capacity and was not successful as a therapeutic (Wolfe et al. 1996; Cavacini et al. 1998).

The monoclonal antibodies 2G12, 2F5, 4E10, and F105 and the immunoglobulin fusion protein PRO 542 have all been probed for efficacy in established HIV infection (Wolfe et al. 1996; Cavacini et al. 1998; Jacobson et al. 2000, 2004a; Armbruster et al. 2004; Nakowitsch et al. 2005; Trkola et al. 2005). These studies were limited in size and produced no conclusive answer on the efficacy or potential of using HIV-specific antibodies as a therapeutic strategy (Gauduin et al. 1997; Mascola et al. 1999; Baba et al. 2000; Mascola et al. 2000; Hofmann-Lehmann et al. 2001; Parren et al. 2001b; Trkola et al. 2005). However, the trials provided important safety, pharmacokinetic, and preliminary antiviral information. All of these antibodies could be delivered at high doses, were tolerated without notable side effects, and had half-lives in the range of other described antibodies in clinical use, ranging from 4.3 to 21.8 days (Table 2). Notably though, despite their in vitro potency, the HIV-specific antibodies have demonstrated no or modest antiviral activity, which was subject to viral escape

Table 2 Serum half-lives of antivirals in humans

Type	Drug	Half-life	Reference
Monoclonal antibodies			
Anti-HIV	2G12	21.8 d	Joos et al. 2006
	4E10	5.5 d	Joos et al. 2006
	2F5	4.3 d	Joos et al. 2006
	F105	13 d	Wolfe et al. 1996
Anti-cell	TNX-355 (anti-CD4)	2.4 d	Jacobson et al. 2004b
	PRO 140 (anti-CCR5)	18 d	Olson et al. 2006
Non-HIV licensed	Synagis (anti-RSV)	20 d	MedImmune 2006
	Hepatitis B-Ig (polyclonal)	21 d	PDR 2006
Peptide/protein inhibitors			
Anti-HIV	CD4-IgG2 (PRO 542)	4.2–3.3 d	Jacobson et. al. 2000
	T-20 (enfuvirtide)	3.8 h	PDR 2006
Small molecule inhibitors			
Anti-HIV (NRTI)	AZT	1.1 h	PDR 2006
	3TC	5–7 h	PDR 2006
	Abacavir	1.5 h	PDR 2006
Anti-HIV (NNRTI)	Efavirenz	52–76 h	PDR 2006
	Nevirapine	45 h	PDR 2006
Anti-HIV (protease inhibitors)	Saquinavir	7 h	PDR 2006
	Ritonavir	3–5 h	PDR 2006
	Indinavir	1.8 h	PDR 2006
Anti-cell (anti-CCR5)	Maraviroc (UK-427)	0.9–2.3 h[a]	Dorr et al. 2005
	Vicriviroc (SCH-D)	3.4 h[b]	Strizki et al. 2005
Non-HIV licensed	Oseltamivir	6–10 h	PDR 2006
	Zanamivir	2.6–5.1 h	PDR 2006
	Acyclovir	2.9 h	PDR 2006

[a] Rat and dog
[b] Rhesus monkey

(Armbruster et al. 2004; Nakowitsch et al. 2005; Trkola et al. 2005). Although comparatively limited, the antiviral effects observed in these studies provided the first direct proof of neutralizing antibody activity in humans. Yet the outcome of these studies raised many questions that will need to be answered in order to drive development of vaccines and antibody therapeutics forward. Central to all is to define why HIV-specific antibodies were not more successful in these trials. The possible reasons are multifaceted. Modes and kinetics of viral transmission may be different in vivo and in vitro. Distribution to relevant sites of viral replication may have not occurred or necessary dose levels may not have been reached. Collectively, these studies suggest that the quantities of HIV-specific antibody needed to control infection in vivo are higher than the in vitro effective doses, perhaps by tenfold or more (Poignard et al. 1999; Veazey et al. 2003; Trkola et al. 2005). Further insight into these issues may be obtained from additional clinical trials, both of HIV-specific antibodies and of antibodies that bind cellular receptors.

2.3 Characteristics of Clinically Tested Anti-HIV Antibodies

2.3.1 MAb 2G12

The antibody 2G12 recognizes a cluster of high mannose carbohydrates of N-linked glycosylated amino acid residues on the immunologically silent face of gp120 (Trkola et al. 1996b; Wyatt et al. 1998a; Sanders et al. 2002; Scanlan et al. 2002) and has broad neutralizing activity in vitro against isolates from subtype B and to a lesser extent also against other subtypes (Burton et al. 1994; Binley et al. 2004; Trkola et al. 2005). The heavy and light chains of 2G12 are not associated in a traditional "Y"-like manner but instead are lying vertically and adjacent to one another. This unique structure provides the antibody with the flexibility to undergo multivalent interactions with the gp120 oligomannose cluster (Calarese et al. 2003). Phase I and phase I/II studies with this antibody have been conducted (Armbruster et al. 2002; Stiegler et al. 2002; Trkola et al. 2005).

2.3.2 The MPER-Specific Antibodies 2F5 and 4E10

The antibodies 2F5 (Muster et al. 1993; Purtscher et al. 1994; Trkola et al. 1995) and 4E10 (Stiegler et al. 2001; Zwick et al. 2001) bind to adjacent linear epitopes on the ectodomain of gp41 in close proximity to the viral membrane (MPER). This MPER of gp41 is accessible to neutralizing antibodies, as recently confirmed by cryoelectron microscopy tomography (Zhu et al. 2006). The core epitopes of 2F5 and 4E10 span amino acids 662–668 (ELDKWAS) and 671–676 [NWF(D/N)IT], respectively. Both antibodies were successfully tested in phase I studies (Armbruster et al. 2002, 2004).

In a recent clinical study we found that a combination of 2G12, 2F5, and 4E10 was able to delay viral rebound in several patients after cessation of successful ART (Trkola et al. 2005). Notably, escape mutant analysis demonstrated that the activity of 2G12 was crucial for the in vivo effect of the neutralizing antibody cocktail in this trial (Trkola et al. 2005).

2.3.3   The Tetravalent CD4-Immunoglobulin Fusion Protein PRO 542

Gp120 binds the most amino-terminal of the four immunoglobulin-like domains of CD4 (Peterson et al. 1988; Arthos et al. 1989; Kwong et al. 1998), and antiviral activity has been demonstrated in CD4-based proteins that incorporate 1, 2, or 4 domains. PRO 542 (CD4-IgG2, Progenics Pharmaceuticals) is a tetravalent CD4-immunoglobulin fusion protein that comprises the D1 and D2 domains of human CD4 genetically fused to the constant domains of both the human IgG2 heavy chain and the κ light chain (Allaway et al. 1995). Compared to monovalent and divalent CD4-based proteins, PRO 542 more broadly and potently neutralizes primary HIV-1 isolates, independent of viral subtype and co-receptor usage, in a variety of preclinical settings (Trkola et al. 1995; Gauduin et al. 1996, 1998; Trkola et al. 1998; Nagashima et al. 2001; Ketas et al. 2003; Rusert et al. 2005). The activity of PRO 542 compares favorably with that of the leading HIV-1 neutralizing mAbs and is preserved also against in vivo viral isolates, primary viruses that have not been cultured in vitro and thus have not acquired higher sensitivity to CD4-based inhibitors (Olson et al. 2003; Beaumont et al. 2004; Jacobson et al. 2004a; Pugach et al. 2004; Shearer et al. 2006). Due to its mechanism of action, PRO 542 can act synergistically with other entry inhibitors, as was demonstrated for enfuvirtide in vitro (Nagashima et al. 2001). Administration of single-dose PRO 542 to treatment-experienced HIV-infected adults at doses ranging from 0.2 to 25 mg/kg (Jacobson et al. 2000; Jacobson et al. 2004a) was generally well tolerated, with no dose-limiting toxicities observed. Mean HIV RNA reductions of approximately 0.5 log10 were observed at the higher dose levels, with a trend toward greater antiviral effects in patients with more advanced disease (Jacobson et al. 2000; Jacobson et al. 2004a). Overall similar results were observed in a pediatric study that examined four weekly doses of 10 mg/kg PRO 542 (Shearer et al. 2000).

3   Targeting the Uninfected Cell

3.1   Modes of Action and Efficacy of Anti-cell Antibodies

The cellular receptors for HIV, CD4 and the co-receptors CCR5 and CXCR4, have proved to be promising targets for entry inhibition, and an array of small molecule inhibitors, antagonists and antibodies targeting these receptors has been developed over the years (Pierson et al. 2003). When used as short-term monotherapy, small-molecule CCR5 antagonists have resulted in 1.5 log10 mean reductions in HIV RNA in patients without target-related toxicities, providing proof-of-concept for targeting host receptors required for entry (Fatkenheuer et al. 2005; Lalezari et al. 2005).

Antibodies to host receptors are attractive from an efficacy perspective, given the immutable nature of the target. For mAbs that bind virus, viral resistance can result from mutations that abrogate antibody binding. However, for mAbs to host receptors, viral resistance requires mutations that circumvent mAb binding. For example, the virus must adapt to no longer require the host receptor, to utilize another site on the receptor, or to utilize the mAb-bound form of the receptor. These escape mechanisms have been described for receptor-targeting drugs and are expected to occur also in response to antibodies (Trkola et al. 2002; Kuhmann et al. 2004; Mosley et al. 2006). Preliminary indications are, however, that these forms of viral resistance typically require multiple mutations and may be slower to develop.

Safety considerations are particularly important for mAbs to host receptors, given the potential to dysregulate immune functions or deplete cells that express the host receptor, as recent incidents have shown (Glass et al. 2006; Suntharalingam et al. 2006). To minimize the risk of unwanted cell destruction, IgG4 mAbs commonly are employed, as these are among the least reactive in terms of Fc effector functions and therefore have a limited potential to eliminate cells by ADCC or complement-dependent cytolysis.

Despite the potential drawbacks, several receptor-targeting antibodies have been developed and have proved safe in clinical application. MAbs to the first immunoglobulin-like domain (D1) of CD4 can inhibit HIV-1 entry, but antibodies tested thus far were shown to induce immune suppression and depletion of CD4+ T cells, and therefore none has been pursued for therapeutic application in HIV infection. Similarly, development of CXCR4 mAbs for HIV-1 therapy has been complicated by the relatively broad tissue distribution of the receptor and its critical role in development and hematopoiesis (Murdoch 2000).

CCR5 and the second immunoglobulin-like domain (D2) of CD4 have proved to provide more viable targets for mAb therapy. A significant fraction of Caucasians lacks a functional CCR5 gene due to naturally occurring mutations (Martinson et al. 1997). The observation that these individuals display no obvious phenotype but are highly resistant to infection by HIV-1 (Liu et al. 1996; Lederman et al. 2006) prompted a focus on developing CCR5-specific drugs and antibodies. Antibodies to CCR5 and CD4 have been identified that do not block the natural activity of these receptors in vitro, increasing the chances that their application in vivo will be safe. As discussed below, both CCR5 and CD4 mAbs are currently progressing through clinical development.

3.2 Characteristics of Clinically Tested Anti-cell Antibodies

3.2.1 CD4 Specific: TNX-355

TNX-355 (Tanox, Houston, TX) is a humanized IgG4 version of the nondepleting murine anti-CD4 mAb, 5A8 (Burkly et al. 1992; Moore et al. 1992; Burkly et al. 1995; Reimann et al. 1997; Reimann et al. 2002). Unlike most HIV-inhibitory

CD4 mAbs, TNX-355 does not block gp120 binding to CD4 but rather blocks a post-attachment event in the entry cascade by binding the D2 domain of CD4. Notably, MAb binding to CD4 does not lead to T cell depletion or immunosuppression (Reimann et al. 1997; Boon et al. 2002; Jacobson et al. 2004b). TNX-355 potently inhibits HIV-1 in a co-receptor-independent manner and demonstrates antiviral synergy with enfuvirtide in vitro (Zhang et al. 2006). In vivo administration of a single dose of TNX-355 to HIV-infected patients was found safe and led, despite the relatively rapid clearance of antibody (serum half-life 2.4 days, Table 2), to a decrease of HIV-1 RNA of 1–1.5 log10 measured 14 days after application (Jacobson et al. 2004b). This durability of the antiviral effect was correlated with prolonged coating of CD4 lymphocytes (Jacobson et al. 2004b). In a current phase II trial, patients were randomized to receive TNX-355 or placebo plus optimized background therapy. TNX-355 was administered intravenously every 2 weeks at doses of 15 mg/kg or 10 mg/kg, with the 10 mg/kg group receiving the first nine doses on a weekly basis. On this trial, 48-week data have been presented (Norris et al. 2006). Both dose levels were well tolerated and demonstrated significant antiviral activity. At 48 weeks, the mean HIV RNA reductions were 0.14, 0.96, and 0.71 log10 for the placebo, TNX-355 10 mg/kg, and TNX-355 15 mg/kg groups, respectively.

3.2.2   CCR5 Specific: PRO 140

PRO 140 (Progenics Pharmaceuticals, Tarrytown, NY) is a humanized IgG4 CCR5 mAb, derived from the murine mAb PA14, which potently and specifically blocks R5 HIV-1 entry in vitro (Olson et al. 1999). PRO 140 and/or PA14 have been shown to broadly inhibit primary R5 HIV-1 isolates independent of genetic subtype (Trkola et al. 2001; Cilliers et al. 2003), HIV-1 disease stage (Rusert et al. 2005), target cell type (Ketas et al. 2003), and resistance to existing antiretrovirals (Olson et al. 2003; Shearer et al. 2006).

Unlike the available small-molecule CCR5 inhibitors (Tagat et al. 2004; Dorr et al. 2005; Takashima et al. 2005; Watson et al. 2005), antiviral concentrations of PRO 140 do not block natural CCR5 function in vitro (Olson et al. 1999). Preliminary studies indicate that PRO 140 is active against viruses that are resistant to small-molecule CCR5 antagonists (Kuhmann et al. 2004; Marozsan et al. 2005) and shows antiviral synergy when combined with small-molecule CCR5 antagonists in vitro (Murga et al. 2006). These complementary properties may reflect the distinct differences in CCR5 binding. Small-molecule CCR5 antagonists bind a hydrophobic pocket formed by the transmembrane helices of CCR5 and inhibit HIV-1 via allosteric mechanisms (Dragic et al. 2000; Tsamis et al. 2003; Nishikawa et al. 2005; Watson et al. 2005), while PRO 140 binds an extracellular epitope on CCR5 and appears to act as a competitive inhibitor (Olson et al. 1999). The synergies and complementary resistance profiles indicate that PRO 140 and small-molecule CCR5 antagonists may represent distinct subclasses of CCR5 inhibitors. Phase I safety testing in healthy individuals showed that concentrations of up to 5 mg/kg PRO 140 were well tolerated in vivo, non immunogenic, and had a half-life of 18 days. Importantly, at the 5 mg/kg dose level, CCR5 lymphocytes were coated with PRO 140 for more than 60 days without cellular depletion (Olson et al. 2006). A phase Ib study is underway to examine the safety, pharmacokinetics, and antiviral effects of single-dose intravenous PRO 140 in individuals with R5 HIV-1 infection.

3.2.3   CCR5 Specific: CCR5mAb004

CCR5mAb004 (Human Genome Sciences, Human Genome Sciences, Rockville, MD) is a fully human IgG4 mAb to CCR5 that inhibits R5 isolates in a subtype-independent manner (Roschke et al. 2004). This mAb inhibits MIP-1β binding to CCR5 and does not induce CCR5 signaling in the absence of chemokine. As expected for an IgG4 antibody, CCR5mAb004 does not mediate ADCC or complement-dependent cytotoxicity (CDC) in vitro (Roschke et al. 2004).

A phase I study was performed in 63 HIV-1 patients with R5 virus who were randomized to receive a single intravenous infusion of placebo or CCR4mAb004 at doses ranging from 0.4 to 40 mg/kg (Lalezari et al. 2006). The antibody was well tolerated overall; however, infusion-related allergic reactions necessitated pre-medication with antihistamines at doses above 2 mg/kg. The serum half-life was 5–8 days, and more than 80% receptor occupancy was observed for 14–28 days with the higher dose cohorts. HIV RNA reductions of 1 log or greater were observed at day 14 in 16 of 29 subjects treated with 8, 20, and 40 mg/kg, with corresponding mean HIV RNA reductions of 0.8–1.0 log10. The findings provide initial proof-of-concept for CCR5 mAb therapy of HIV-1 infection.

4   Potential Advantages of Therapeutic Antibodies

Several factors make antibodies an attractive class of molecules for HIV-1 therapy. The most valuable feature of antibodies lies in their intrinsic nature: the immune system engineers them to recognize their target with high specificity and affinity. Therapy of various conditions with polyclonal and, more recently, monoclonal antibodies has been performed for decades and proved to be generally safe (Sawyer 2000; Zeitlin et al. 2000; Reichert et al. 2005). Metabolic side effects, as observed with many small-molecule drugs, are generally not seen due to their predictable catabolism into naturally occurring amino acids. Additionally, unlike small-molecule drugs, mAbs do not passively diffuse across cellular membranes, reducing their potential for metabolic and other nontarget toxicities. Therefore, mAb therapy could be expected to offer an improved or at least nonoverlapping side effect profile compared to existing antiretrovirals. Unless selected to react with host moieties, cross reactivity with host antigens is rare and usually can be excluded during preclinical development. Antigenicity of human or humanized antibodies is commonly low, allowing continued, high-dose application. Notably also, half-lives of humanized and human antibodies are, in general, considerably higher than those of small-molecule inhibitors (Table 2). All HIV-1 drugs currently in use have to be administered once to several times per day to maintain therapeutic levels. If drug adherence is not strict, resistant viral strains evolve rapidly. Antibodies, if provided as a component of ART, could be of benefit in this regard. Due to their typically long serum half-lives, they could enable infrequent dosing that does not require daily vigilance from the patient.

In addition, several of the currently used antiretrovirals (protease inhibitors, nonnucleoside reverse transcriptase inhibitors) are substrates or inducers of cyto-chrome P450 enzymes and thus can substantially perturb metabolic pathways (Cressey et al. 2006). Several of these drugs are indeed associated with signifi-cant drug–drug interactions between antiretroviral drugs used in combination and between additional medications the patient may require for other conditions (de Maat et al. 2003; Winston et al. 2005). Antibodies could in this setting be advan-tageous as they are not metabolized by cytochrome P450 enzymes, and thus could simplify the selection of combination treatment regimens. Lastly, mAbs may efficiently block protein–protein interactions or other targets that are not readily drugable with small molecules. For example, there presently are no licensed small-molecule drugs that target the CD4, CCR5, gp120, and gp41 epitopes recognized by the mAbs TNX-355, PRO 140, 2G12, and 2F5/4E10, respectively. Such agents can be expected to inhibit viruses that are broadly resistant to the available antiretroviral therapies.

A potential drawback of antibody therapeutics is the cost of production. However, progressive advances in cell engineering and bioreactor operation have enabled these products to be manufactured efficiently in mammalian cells, and cur-rent processes often yield multi-gram per liter expression in chemically defined medium (Butler 2005). A further limitation of mAb therapy is the need to deliver the product by injection for systemic application. However, infrequent, self-administered injections may provide an attractive alternative to daily pill regimens for many patients, and needleless delivery devices offer a means to further improve patient acceptance.

Theoretically, various clinical settings can be envisioned where a nontoxic, long-acting therapy could be beneficial (Box 1 and Table 1). Like other investiga-tional drugs for HIV, antibody therapeutics are being developed for treatment of HIV-1 in combination with other antiretroviral agents as a component of ART, and antibody therapeutics have the potential to offer new treatment classes and addi-tions to the armamentarium of HIV drugs. As with any new drug for HIV, antibody therapeutics initially may find the greatest use in treatment-experienced patients with fewer treatment options. However, as clinical experience increases, use in earlier stage patients could be expected to increase as warranted by the safety and efficacy profile of the molecule.

Where available, ART of infected mothers has dramatically lowered transmis-sion rates (De Cock et al. 2000; UNAIDS 2006). Antibody therapeutics have been suggested as a potential adjunct to ART therapy in MTCT as they could extend protection throughout the breast-feeding period. Passive immunization to prevent MTCT has been considered for a long time and may provide an option, as not all

HIV drugs are approved for pediatric use. In untreated mothers, transmission rates before and during birth are high and an almost equally high proportion of infant infections is thought to be acquired through breast feeding (De Cock et al. 2000; UNAIDS 2006). While bearing the risk of HIV infection, abstaining breast feeding in these settings is problematic as it can lead to malnutrition of the newborn and increased mortality due to other infections. Likewise, passive immunization or combination of active and passive immunization, as successfully employed against HBV infection (Kabir et al. 2006), could help to reduce transmission postpartum.

Use of antibodies as topical microbicides may equally come in reach, since various means for the controlled release of local delivery of therapeutic antibodies have been developed and are in clinical use in other settings (Grainger 2004). As mentioned already, a potentially dramatic impact of antibodies in therapy could be envisioned if these antibodies target and destruct infected cells through activation of effector functions or by delivering immunotoxins.

5   Future Perspectives

The current generation of antibodies has provided initial insights into modes and potential of antibody therapeutics, and the accumulated knowledge provides a solid basis for further development (Box 1). Importantly, all antibodies tested to date have shown favorable safety profiles in man. In addition, compared to HIV-specific antibodies, antibodies to host receptors have shown more promising antiviral activity. CD4 and CCR5 mAbs currently are progressing through controlled clinical trials, and the results undoubtedly will add to our understanding of the potential role of antibody therapeutics in HIV.

To date, the antiviral effects seen for HIV-specific antibodies have been modest. For these to be effective, we will need novel antibodies with enhanced features (Box 1). Our best-characterized neutralizing antibodies, despite their comparatively broad activity, preferentially recognize the subtype B isolates against which they originated. Novel scaffolds for epitope presentation or envelope structure mimetics that overcome the limitations of previous antigens used for vaccination are under development and are anticipated to foster the isolation of antibodies with novel specificities.

Impressive strides in antibody engineering have been made in recent years. Antibody affinity can be enhanced through methods of directed evolution (Carter 2006; Luginbuhl et al. 2006). Increased affinity may reduce the amount of antibody required and thereby improve the affordability and delivery of these products. For example, high potency may be a requirement for self-administrable, sub-cutaneous formulations of antibody. Directed evolution of antibody affinity is especially attractive for antibodies to invariant host receptors. Antibody engineering has been successful in enhancing effector functions by more than two orders of magnitude (Umana et al. 1999; Lazar et al. 2006) and in eliminating

Box 1 Future perspective: What is needed to drive development of antibody therapeutics forward:

- Discover novel antibodies
    - Define new isolation and screening methods for HIV-specific mAbs
    - Define new target epitopes
    - Isolate antibodies that neutralize divergent genetic subtypes with high activity
    - Define modes of action in vivo (neutralization versus effector functions)
    - Define modes of transmission in vivo (free virus versus cell-to-cell)
- Consider in vitro engineering of antibody characteristics
    - Improve antiviral activity
    - Enhance activation of effector functions
    - Increase stability and half-life
    - Use immunotoxin design.
- Probe antibody combinations
    - Use multiple epitope specificities
    - Use neutralizing and effector function-inducing antibodies
    - Use antiviral and anti-cell antibodies
- Improve production, formulation, and delivery
    - Develop self-administered formulations
    - Allow for needleless delivery
    - Develop controlled release and local delivery systems for topical and systemic application
- Probe clinical efficacy in relevant settings
    - Combine with optimized background ART
    - Use in prevention of mother-to-child transmission
    - Include a microbicide

residual effector functions, if desired (Hsu et al. 1999). Additional modifications can improve the serum half-life of antibodies and thereby reduce dose levels and intervals (Hinton et al. 2006). This tool-chest of technologies can be exploited to potentially optimize the efficacy, safety, and convenience of antibody therapies for HIV.

6   Conclusion

Developing safe and effective antibody therapies for HIV infection is rich in challenge and opportunity. Any single approach to battling HIV bears a high risk of failure; however, this risk is spread among the diversity of targets and modalities for antibody therapy. Recent clinical trials have provided initial optimism for antibodies to host receptors and a foundation for further studies. The prospects for HIV-specific antibodies as therapeutic agents are less clear at present. But can we afford not to try this approach? At minimum, there is much that HIV-specific antibodies can teach us about vaccine design. By fully exploring the possibilities of antibody therapies for HIV, we might end up with both better vaccines and new therapeutics.

Acknowledgements We thank Huldrych F. Günthard for critical reading of the manuscript and helpful discussions. Support was provided by the Swiss National Science Foundation (grant No. PP00B-102647 to A.T.) and by NIH (grant No. AI066329 to W.C.O). A.T. is an Elizabeth Glaser Scientist supported by the Elizabeth Glaser Pediatric AIDS Foundation.

References

Aasa-Chapman MM, Holuigue S, Aubin K, Wong M, Jones NA, Cornforth D, Pellegrino P, Newton P, Williams I, Borrow P, McKnight A (2005) Detection of antibody-dependent complement-mediated inactivation of both autologous and heterologous virus in primary human immunodeficiency virus type 1 infection. J Virol 79:2823–2830

Ahmad A, Yao XA, Tanner JE, Cohen E, Menezes J (1994) Surface expression of the HIV-1 envelope proteins in env gene-transfected CD4-positive human T cell clones: characterization and killing by an antibody-dependent cellular cytotoxic mechanism. J Acquir Immune Defic Syndr 7:789–798

Ahmad R, Sindhu ST, Toma E, Morisset R, Vincelette J, Menezes J, Ahmad A (2001) Evidence for a correlation between antibody-dependent cellular cytotoxicity-mediating anti-HIV-1 antibodies and prognostic predictors of HIV infection. J Clin Immunol 21:227–233

Alkhatib G, Combadiere C, Broder CC, Feng Y, Kennedy PE, Murphy PM, Berger EA (1996) CC CKR5: a RANTES, MIP-1alpha, MIP-1beta receptor as a fusion cofactor for macrophage-tropic HIV-1. Science 272:1955–1958

Allaway GP, Davis-Bruno KL, Beaudry GA, Garcia EB, Wong EL, Ryder AM, Hasel KW, Gauduin MC, Koup RA, McDougal JS, et al (1995) Expression and characterization of CD4-IgG2, a novel heterotetramer that neutralizes primary HIV type 1 isolates. AIDS Res Hum Retroviruses 11:533–539

Armbruster C, Stiegler GM, Vcelar BA, Jager W, Michael NL, Vetter N, Katinger HW (2002) A phase I trial with two human monoclonal antibodies (hMAb 2F5, 2G12) against HIV-1. Aids 16:227–233

Armbruster C, Stiegler GM, Vcelar BA, Jager W, Koller U, Jilch R, Ammann CG, Pruenster M, Stoiber H, Katinger HW (2004) Passive immunization with the anti-HIV-1 human monoclonal antibody (hMAb) 4E10 and the hMAb combination 4E10/2F5/2G12. J Antimicrob Chemother 54:915–920

Arrighi JF, Pion M, Garcia E, Escola JM, van Kooyk Y, Geijtenbeek TB, Piguet V (2004) DC-SIGN-mediated infectious synapse formation enhances X4 HIV-1 transmission from dendritic cells to T cells. J Exp Med 200:1279–1288

Arthos J, Deen KC, Chaikin MA, Fornwald JA, Sathe G, Sattentau QJ, Clapham PR, Weiss RA, McDougal JS, Pietropaolo C, et al (1989) Identification of the residues in human CD4 critical for the binding of HIV. Cell 57:469–481

Azzam R, Kedzierska K, Leeansyah E, Chan H, Doischer D, Gorry PR, Cunningham AL, Crowe SM, Jaworowski A (2006) Impaired complement-mediated phagocytosis by HIV type-1-infected human monocyte-derived macrophages involves a cAMP-dependent mechanism. AIDS Res Hum Retroviruses 22:619–629

Baba TW, Liska V, Hofmann-Lehmann R, Vlasak J, Xu W, Ayehunie S, Cavacini LA, Posner MR, Katinger H, Stiegler G, Bernacky BJ, Rizvi TA, Schmidt R, Hill LR, Keeling ME, Lu Y, Wright JE, Chou TC, Ruprecht RM (2000) Human neutralizing monoclonal antibodies of the IgG1 subtype protect against mucosal simian-human immunodeficiency virus infection. Nat Med 6:200–206

Bangham CR (2003) The immune control and cell-to-cell spread of human T-lymphotropic virus type 1. J Gen Virol 84:3177–3189

Banks ND, Kinsey N, Clements J, Hildreth JE (2002) Sustained antibody-dependent cell-mediated cytotoxicity (ADCC) in SIV-infected macaques correlates with delayed progression to AIDS. AIDS Res Hum Retroviruses 18:1197–1205

Baum LL, Cassutt KJ, Knigge K, Khattri R, Margolick J, Rinaldo C, Kleeberger CA, Nishanian P, Henrard DR, Phair J (1996) HIV-1 gp120-specific antibody-dependent cell-mediated cytotoxicity correlates with rate of disease progression. J Immunol 157:2168–2173

Beaumont T, Quakkelaar E, van Nuenen A, Pantophlet R, Schuitemaker H (2004) Increased sensitivity to CD4 binding site-directed neutralization following in vitro propagation on primary lymphocytes of a neutralization-resistant human immunodeficiency virus IIIB strain isolated from an accidentally infected laboratory worker. J Virol 78:5651–5657

Belo M, Yagello M, Girard M, Greenlee R, Deslandres A, Barre-Sinoussi F, Gluckman JC (1991) Antibody-dependent cellular cytotoxicity against HIV-1 in sera of immunized chimpanzees. AIDS 5:169–176

Bender BS, Davidson BL, Kline R, Brown C, Quinn TC (1988) Role of the mononuclear phagocyte system in the immunopathogenesis of human immunodeficiency virus infection and the acquired immunodeficiency syndrome. Rev Infect Dis 10:1142–1154

Binley JM, Wrin T, Korber B, Zwick MB, Wang M, Chappey C, Stiegler G, Kunert R, Zolla-Pazner S, Katinger H, Petropoulos CJ, Burton DR (2004) Comprehensive cross-clade neutralization analysis of a panel of anti-human immunodeficiency virus type 1 monoclonal antibodies. J Virol 78: 13232–13252

Blue CE, Spiller OB, Blackbourn DJ (2004) The relevance of complement to virus biology. Virology 319:176–184

Boon L, Holland B, Gordon W, Liu P, Shiau F, Shanahan W, Reimann KA, Fung M (2002) Development of anti-CD4 MAb hu5A8 for treatment of HIV-1 infection: preclinical assessment in non-human primates. Toxicology 172:191–203

Brekke OH, Sandlie I (2003) Therapeutic antibodies for human diseases at the dawn of the twenty-first century. Nat Rev Drug Discov 2:52–62

Broliden K, Sievers E, Tovo PA, Moschese V, Scarlatti G, Broliden PA, Fundaro C, Rossi P (1993) Antibody-dependent cellular cytotoxicity and neutralizing activity in sera of HIV-1-infected mothers and their children. Clin Exp Immunol 93:56–64

Broliden K, Hinkula J, Tolfvenstam T, Niphuis H, Heeney J (1996) Antibody-dependent cellular cytotoxicity to clinical isolates of HIV-1 and SIVcpz: comparison of human and chimpanzees. AIDS 10:1199–1204

Bross PF, Beitz J, Chen G, Chen XH, Duffy E, Kieffer L, Roy S, Sridhara R, Rahman A, Williams G, Pazdur R (2001) Approval summary: gemtuzumab ozogamicin in relapsed acute myeloid leukemia. Clin Cancer Res 7:1490–1496

Burkly L, Mulrey N, Blumenthal R, Dimitrov DS (1995) Synergistic inhibition of human immunodeficiency virus type 1 envelope glycoprotein-mediated cell fusion and infection by an antibody to CD4 domain 2 in combination with anti-gp120 antibodies. J Virol 69: 4267–4273

Burkly LC, Olson D, Shapiro R, Winkler G, Rosa JJ, Thomas DW, Williams C, Chisholm P (1992) Inhibition of HIV infection by a novel CD4 domain 2-specific monoclonal antibody. Dissecting the basis for its inhibitory effect on HIV-induced cell fusion. J Immunol 149:1779–1787

Burton DR, Pyati J, Koduri R, Sharp SJ, Thornton GB, Parren PW, Sawyer LS, Hendry RM, Dunlop N, Nara PL, et al (1994) Efficient neutralization of primary isolates of HIV-1 by a recombinant human monoclonal antibody. Science 266:1024–1027

Butler M (2005) Animal cell cultures: recent achievements and perspectives in the production of biopharmaceuticals. Appl Microbiol Biotechnol 68:283–291

Calarese DA, Scanlan CN, Zwick MB, Deechongkit S, Mimura Y, Kunert R, Zhu P, Wormald MR, Stanfield RL, Roux KH, Kelly JW, Rudd PM, Dwek RA, Katinger H, Burton DR, Wilson IA (2003) Antibody domain exchange is an immunological solution to carbohydrate cluster recognition. Science 300:2065–2071

Carter PJ (2006) Potent antibody therapeutics by design. Nat Rev Immunol 6:343–357

Cavacini LA, Samore MH, Gambertoglio J, Jackson B, Duval M, Wisnewski A, Hammer S, Koziel C, Trapnell C, Posner MR (1998) Phase I study of a human monoclonal antibody directed against the CD4-binding site of HIV type 1 glycoprotein 120. AIDS Res Hum Retroviruses 14:545–550

Cianci C, Genovesi EV, Lamb L, Medina I, Yang Z, Zadjura L, Yang H, D'Arienzo C, Sin N, Yu KL, Combrink K, Li Z, Colonno R, Meanwell N, Clark J, Krystal M (2004) Oral efficacy of a respiratory syncytial virus inhibitor in rodent models of infection. Antimicrob Agents Chemother 48:2448–2454

Cilliers T, Nhlapo J, Coetzer M, Orlovic D, Ketas T, Olson WC, Moore JP, Trkola A, Morris L (2003) The CCR5 and CXCR4 coreceptors are both used by human immunodeficiency virus type 1 primary isolates from subtype C. J Virol 77:4449–4456

Cressey TR, Lallemant M (2006) Pharmacogenetics of antiretroviral drugs for the treatment of HIV-infected patients: an update. Infect Genet Evol 7:333–342

Davey RT Jr, Boenning CM, Herpin BR, Batts DH, Metcalf JA, Wathen L, Cox SR, Polis MA, Kovacs JA, Falloon J, et al (1994) Use of recombinant soluble CD4 Pseudomonas exotoxin, a novel immunotoxin, for treatment of persons infected with human immunodeficiency virus. J Infect Dis 170:1180–1188

De Cock KM, Fowler MG, Mercier E, de Vincenzi I, Saba J, Hoff E, Alnwick DJ, Rogers M, Shaffer N (2000) Prevention of mother-to-child HIV transmission in resource-poor countries: translating research into policy and practice. JAMA 283:1175–1182

de Maat MM, Ekhart GC, Huitema AD, Koks CH, Mulder JW, Beijnen JH (2003) Drug interactions between antiretroviral drugs and comedicated agents. Clin Pharmacokinet 42:223

Decker JM, Bibollet-Ruche F, Wei X, Wang S, Levy DN, Wang W, Delaporte E, Peeters M, Derdeyn CA, Allen S, Hunter E, Saag MS, Hoxie JA, Hahn BH, Kwong PD, Robinson JE, Shaw GM (2005) Antigenic conservation and immunogenicity of the HIV coreceptor binding site. J Exp Med 201:1407–1419

Deng H, Liu R, Ellmeier W, Choe S, Unutmaz D, Burkhart M, Di Marzio P, Marmon S, Sutton RE, Hill CM, Davis CB, Peiper SC, Schall TJ, Littman DR, Landau NR (1996) Identification of a major co-receptor for primary isolates of HIV-1. Nature 381:661–666

Ding WD, Mitsner B, Krishnamurthy G, Aulabaugh A, Hess CD, Zaccardi J, Cutler M, Feld B, Gazumyan A, Raifeld Y, Nikitenko A, Lang SA, Gluzman Y, O'Hara B, Ellestad GA (1998) Novel and specific respiratory syncytial virus inhibitors that target virus fusion. J Med Chem 41:2671–2675

Dorr P, Westby M, Dobbs S, Griffin P, Irvine B, Macartney M, Mori J, Rickett G, Smith-Burchnell C, Napier C, Webster R, Armour D, Price D, Stammen B, Wood A, Perros M (2005) Maraviroc (UK-427,857), a potent, orally bioavailable, and selective small-molecule inhibitor of chemokine receptor CCR5 with broad-spectrum anti-human immunodeficiency virus type 1 activity. Antimicrob Agents Chemother 49:4721–4732

Dragic T, Litwin V, Allaway GP, Martin SR, Huang Y, Nagashima KA, Cayanan C, Maddon PJ, Koup RA, Moore JP, Paxton WA (1996) HIV-1 entry into CD4+ cells is mediated by the chemokine receptor CC-CKR-5. Nature 381:667–673

Dragic T, Trkola A, Thompson DA, Cormier EG, Kajumo FA, Maxwell E, Lin SW, Ying W, Smith SO, Sakmar TP, Moore JP (2000) A binding pocket for a small molecule inhibitor of HIV-1 entry within the transmembrane helices of CCR5. Proc Natl Acad Sci USA 97:5639–5644

Fatkenheuer G, Pozniak AL, Johnson MA, Plettenberg A, Staszewski S, Hoepelman AI, Saag MS, Goebel FD, Rockstroh JK, Dezube BJ, Jenkins TM, Medhurst C, Sullivan JF, Ridgway C,

Abel S, James IT, Youle M, van der Ryst E (2005) Efficacy of short-term monotherapy with maraviroc, a new CCR5 antagonist, in patients infected with HIV-1. Nat Med 11:1170–1172

Feng Y, Broder CC, Kennedy PE, Berger EA (1996) HIV-1 entry cofactor: functional cDNA cloning of a seven-transmembrane, G protein-coupled receptor. Science 272:872–877

Ferrari G, Place CA, Ahearne PM, Nigida SM Jr, Arthur LO, Bolognesi DP, Weinhold KJ (1994) Comparison of anti-HIV-1 ADCC reactivities in infected humans and chimpanzees. J Acquir Immune Defic Syndr 7:325–331

Finzi D, Hermankova M, Pierson T, Carruth LM, Buck C, Chaisson RE, Quinn TC, Chadwick K, Margolick J, Brookmeyer R, Gallant J, Markowitz M, Ho DD, Richman DD, Siliciano RF (1997) Identification of a reservoir for HIV-1 in patients on highly active antiretroviral therapy. Science 278:1295–1300

Finzi D, Blankson J, Siliciano JD, Margolick JB, Chadwick K, Pierson T, Smith K, Lisziewicz J, Lori F, Flexner C, Quinn TC, Chaisson RE, Rosenberg E, Walker B, Gange S, Gallant J, Siliciano RF (1999) Latent infection of CD4+ T cells provides a mechanism for lifelong persistence of HIV-1, even in patients on effective combination therapy. Nat Med 5:512–517

Forthal DN, Landucci G, Daar ES (2001) Antibody from patients with acute human immunodeficiency virus (HIV) infection inhibits primary strains of HIV type 1 in the presence of natural-killer effector cells. J Virol 75:6953–6961

Gauduin MC, Allaway GP, Maddon PJ, Barbas CF, 3rd Burton DR, Koup RA (1996) Effective ex vivo neutralization of human immunodeficiency virus type 1 in plasma by recombinant immunoglobulin molecules. J Virol 70:2586–2592

Gauduin MC, Parren PW, Weir R, Barbas CF, Burton DR, Koup RA (1997) Passive immunization with a human monoclonal antibody protects hu-PBL-SCID mice against challenge by primary isolates of HIV-1. Nat Med 3:1389–1393

Gauduin MC, Allaway GP, Olson WC, Weir R, Maddon PJ, Koup RA (1998) CD4-immunoglobulin G2 protects Hu-PBL-SCID mice against challenge by primary human immunodeficiency virus type 1 isolates. J Virol 72:3475–3478

Glass WG, McDermott DH, Lim JK, Lekhong S, Yu SF, Frank WA, Pape J, Cheshier RC, Murphy PM (2006) CCR5 deficiency increases risk of symptomatic West Nile virus infection. J Exp Med 203:35–40

Golay J, Manganini M, Facchinetti V, Gramigna R, Broady R, Borleri G, Rambaldi A, Introna M (2003) Rituximab-mediated antibody-dependent cellular cytotoxicity against neoplastic B cells is stimulated strongly by interleukin-2. Haematologica 88:1002–1012

Grainger DW (2004) Controlled-release and local delivery of therapeutic antibodies. Expert Opin Biol Ther 4:1029–1044

Gulick RM, Ribaudo HJ, Shikuma CM, Lalama C, Schackman BR, Meyer WA, 3rd Acosta EP, Schouten J, Squires KE, Pilcher CD, Murphy RL, Koletar SL, Carlson M, Reichman RC, Bastow B, Klingman KL, Kuritzkes DR (2006) Three- vs four-drug antiretroviral regimens for the initial treatment of HIV-1 infection: a randomized controlled trial. JAMA 296:769–781

Gunthard HF, Gowland PL, Schupbach J, Fung MS, Boni J, Liou RS, Chang NT, Grob P, Graepel P, Braun DG, et al (1994) A phase I/IIA clinical study with a chimeric mouse-human monoclonal antibody to the V3 loop of human immunodeficiency virus type 1 gp120. J Infect Dis 170: 1384–1393

Haigwood NL, Montefiori DC, Sutton WF, McClure J, Watson AJ, Voss G, Hirsch VM, Richardson BA, Letvin NL, Hu SL, Johnson PR (2004) Passive immunotherapy in simian immunodeficiency virus-infected macaques accelerates the development of neutralizing antibodies. J Virol 78:5983–5995

Hangartner L, Zellweger RM, Giobbi M, Weber J, Eschli B, McCoy KD, Harris N, Recher M, Zinkernagel RM, Hengartner H (2006) Nonneutralizing antibodies binding to the surface glycoprotein of lymphocytic choriomeningitis virus reduce early virus spread. J Exp Med 203:2033–2042

Haynes BF, Fleming J, St Clair EW, Katinger H, Stiegler G, Kunert R, Robinson J, Scearce RM, Plonk K, Staats HF, Ortel TL, Liao HX, Alam SM (2005) Cardiolipin polyspecific autoreactivity in two broadly neutralizing HIV-1 antibodies. Science 308:1906–1908

Hinton PR, Xiong JM, Johlfs MG, Tang MT, Keller S, Tsurushita N (2006) An engineered human IgG1 antibody with longer serum half-life. J Immunol 176:346–356

Hober D, Jewett A, Bonavida B (1995) Lysis of uninfected HIV-1 gp120-coated peripheral blood-derived T lymphocytes by monocyte-mediated antibody-dependent cellular cytotoxicity. FEMS Immunol Med Microbiol 10:83–91

Hofmann-Lehmann R, Vlasak J, Rasmussen RA, Smith BA, Baba TW, Liska V, Ferrantelli F, Montefiori DC, McClure HM, Anderson DC, Bernacky BJ, Rizvi TA, Schmidt R, Hill LR, Keeling ME, Katinger H, Stiegler G, Cavacini LA, Posner MR, Chou TC, Andersen J, Ruprecht RM (2001) Postnatal passive immunization of neonatal macaques with a triple combination of human monoclonal antibodies against oral simian-human immunodeficiency virus challenge. J Virol 75:7470–7480

Hsu DH, Shi JD, Homola M, Rowell TJ, Moran J, Levitt D, Druilhet B, Chinn J, Bullock C, Klingbeil C (1999) A humanized anti-CD3 antibody, HuM291, with low mitogenic activity, mediates complete and reversible T-cell depletion in chimpanzees. Transplantation 68:545–554

Huber M, Fischer M, Misselwitz B, Manrique A, Kuster H, Niederöst B, Weber R, Von Wyl V, Günthard HF, Trkola A (2006) Complement lysis activity in autologous plasma is associated with lower viral loads during the acute phase of HIV-1 infection. PLoS Med 3:e440

Huntley CC, Weiss WJ, Gazumyan A, Buklan A, Feld B, Hu W, Jones TR, Murphy T, Nikitenko AA, O'Hara B, Prince G, Quartuccio S, Raifeld YE, Wyde P, O'Connell JF (2002) RFI-641, a potent respiratory syncytial virus inhibitor. Antimicrob Agents Chemother 46:841–847

Jacobson JM, Lowy I, Fletcher CV, O'Neill TJ, Tran DN, Ketas TJ, Trkola A, Klotman ME, Maddon PJ, Olson WC, Israel RJ (2000) Single-dose safety, pharmacology, and antiviral activity of the human immunodeficiency virus (HIV) type 1 entry inhibitor PRO 542 in HIV-infected adults. J Infect Dis 182:326–329

Jacobson JM, Israel RJ, Lowy I, Ostrow NA, Vassilatos LS, Barish M, Tran DN, Sullivan BM, Ketas TJ, O'Neill TJ, Nagashima KA, Huang W, Petropoulos CJ, Moore JP, Maddon PJ, Olson WC (2004a) Treatment of advanced human immunodeficiency virus type 1 disease with the viral entry inhibitor PRO 542. Antimicrob Agents Chemother 48:423–429

Jacobson JM, Kuritzkes E, Godofsky E, DeJesus E, Lewis S, Jackson J, Frazier K, Fagan EA, Shanahan WR (2004b) Phase 1b study of the anti-CD4 monoclonal antibody TNX-355 in HIV-1-infected subjects: safety and antiretroviral activity of multiple doses. 11th Conference on Retroviruses and Opportunistic Infections, San Francisco, CA

Johansson S, Goldenberg DM, Griffiths GL, Wahren B, Hinkula J (2006) Elimination of HIV-1 infection by treatment with a doxorubicin-conjugated anti-envelope antibody. Aids 20:1911–1915

Jolly C, Sattentau QJ (2004b) Retroviral spread by induction of virological synapses. Traffic 5:643–650

Jolly C, Kashefi K, Hollinshead M, Sattentau QJ (2004a) HIV-1 cell to cell transfer across an Env-induced, actin-dependent synapse. J Exp Med 199:283–293

Joos B, Trkola A, Kuster H, Aceto L, Fischer M, Stiegler G, Armbruster C, Vcelar B, Katinger H, Gunthard HF (2006) Long-term multiple-dose pharmacokinetics of human monoclonal antibodies (MAbs) against human immunodeficiency virus type 1 envelope gp120 (MAb 2G12) and gp41 (MAbs 4E10 and 2F5). Antimicrob Agents Chemother 50:1773–1779

Kabir A, Alavian SM, Ahanchi N, Malekzadeh R (2006) Combined passive and active immunoprophylaxis for preventing perinatal transmission of the hepatitis B virus in infants born to HBsAg positive mothers in comparison with vaccine alone. Hepatol Res 36:265–271

Keller MA, Stiehm ER (2000) Passive immunity in prevention and treatment of infectious diseases. Clin Microbiol Rev 13:602–614

Kennedy PE, Bera TK, Wang QC, Gallo M, Wagner W, Lewis MG, Berger EA, Pastan I (2006) Anti-HIV-1 immunotoxin 3B3(Fv)-PE38: enhanced potency against clinical isolates in human PMBCs and macrophages, and negligible hepatotoxicity in macaques. J Leukoc Biol 80:1175–1182

Ketas TJ, Frank I, Klasse PJ, Sullivan BM, Gardner JP, Spenlehauer C, Nesin M, Olson WC, Moore JP, Pope M (2003) Human immunodeficiency virus type 1 attachment, coreceptor, and fusion inhibitors are active against both direct and trans infection of primary cells. J Virol 77:2762–2767

Klasse PJ, Moore JP (2004) Is there enough gp120 in the body fluids of HIV-1-infected individuals to have biologically significant effects? Virology 323:1–8

Klasse PJ, Sattentau QJ (2002) Occupancy and mechanism in antibody-mediated neutralization of animal viruses. J Gen Virol 83:2091–2108

Kuhmann SE, Pugach P, Kunstman KJ, Taylor J, Stanfield RL, Snyder A, Strizki JM, Riley J, Baroudy BM, Wilson IA, Korber BT, Wolinsky SM, Moore JP (2004) Genetic and phenotypic analyses of human immunodeficiency virus type 1 escape from a small-molecule CCR5 inhibitor. J Virol 78:2790–2807

Kulkosky J, Bray S (2006) HAART-persistent HIV-1 latent reservoirs: their origin, mechanisms of stability and potential strategies for eradication. Curr HIV Res 4:199–208

Kwong PD, Wyatt R, Robinson J, Sweet RW, Sodroski J, Hendrickson WA (1998) Structure of an HIV gp120 envelope glycoprotein in complex with the CD4 receptor and a neutralizing human antibody. Nature 393:648–659

Lalezari J, Thompson M, Kumar P, Piliero P, Davey R, Patterson K, Shachoy-Clark A, Adkison K, Demarest J, Lou Y, Berrey M, Piscitelli S (2005) Antiviral activity and safety of 873140, a novel CCR5 antagonist, during short-term monotherapy in HIV-infected adults. Aids 19:1443–1448

Lalezari J, Lederman M, Yadavalli G, Para M, DeJesus E, Searle J, Cai W, Roschke V, Zhong J, Hicks C, Freimuth W, Subramanian M (2006) A phase 1, dose-escalation, placebo controlled study of a fully human monoclonal antibody (CCR5mAb004) against CCR5 in patients with CCR5 tropic HIV-1 infection. 46th International Conference on Antimicrobial Agents and Chemotherapy, San Francisco, CA

Lazar GA, Dang W, Karki S, Vafa O, Peng JS, Hyun L, Chan C, Chung HS, Eivazi A, Yoder SC, Vielmetter J, Carmichael DF, Hayes RJ, Dahiyat BI (2006) Engineered antibody Fc variants with enhanced effector function. Proc Natl Acad Sci U S A 103:4005–4010

Lederman MM, Penn-Nicholson A, Cho M, Mosier D (2006) Biology of CCR5 and its role in HIV infection and treatment. JAMA 296:815–826

Liu R, Paxton WA, Choe S, Ceradini D, Martin SR, Horuk R, MacDonald ME, Stuhlmann H, Koup RA, Landau NR (1996) Homozygous defect in HIV-1 coreceptor accounts for resistance of some multiply-exposed individuals to HIV-1 infection. Cell 86:367–377

Lueders KK, De Rosa SC, Valentin A, Pavlakis GN, Roederer M, Hamer DH (2004) A potent anti-HIV immunotoxin blocks spreading infection by primary HIV type 1 isolates in multiple cell types. AIDS Res Hum Retroviruses 20:145–150

Luginbuhl B, Kanyo Z, Jones RM, Fletterick RJ, Prusiner SB, Cohen FE, Williamson RA, Burton DR, Pluckthun A (2006) Directed evolution of an anti-prion protein scFv fragment to an affinity of 1 pM and its structural interpretation. J Mol Biol 363:75–97

Lyerly HK, Reed DL, Matthews TJ, Langlois AJ, Ahearne PA, Petteway SR Jr, Weinhold KJ (1987) Anti-GP 120 antibodies from HIV seropositive individuals mediate broadly reactive anti-HIV ADCC. AIDS Res Hum Retroviruses 3:409–422

Maddon PJ, Dalgleish AG, McDougal JS, Clapham PR, Weiss RA, Axel R (1986) The T4 gene encodes the AIDS virus receptor and is expressed in the immune system and the brain. Cell 47:333–348

Marozsan AJ, Kuhmann SE, Morgan T, Herrera C, Rivera-Troche E, Xu S, Baroudy BM, Strizki J, Moore JP (2005) Generation and properties of a human immunodeficiency virus type 1 isolate resistant to the small molecule CCR5 inhibitor, SCH-417690 (SCH-D). Virology 338:182–199

Marschang P, Gurtler L, Totsch M, Thielens NM, Arlaud GJ, Hittmair A, Katinger H, Dierich MP (1993) HIV-1 and HIV-2 isolates differ in their ability to activate the complement system on the surface of infected cells. Aids 7:903–910

Marschang P, Sodroski J, Wurzner R, Dierich MP (1995) Decay-accelerating factor (CD55) protects human immunodeficiency virus type 1 from inactivation by human complement. Eur J Immunol 25:285–290

Martinson JJ, Chapman NH, Rees DC, Liu YT, Clegg JB (1997) Global distribution of the CCR5 gene 32-basepair deletion. Nat Genet 16:100

Mascola JR, Lewis MG, Stiegler G, Harris D, VanCott TC, Hayes D, Louder MK, Brown CR, Sapan CV, Frankel SS, Lu Y, Robb ML, Katinger H, Birx DL (1999) Protection of macaques against pathogenic simian/human immunodeficiency virus 89. 6PD by passive transfer of neutralizing antibodies. J Virol 73:4009–4018

Mascola JR, Stiegler G, VanCott TC, Katinger H, Carpenter CB, Hanson CE, Beary H, Hayes D, Frankel SS, Birx DL, Lewis MG (2000) Protection of macaques against vaginal transmission of a pathogenic HIV-1/SIV chimeric virus by passive infusion of neutralizing antibodies. Nat Med 6:207–210

McDonald D, Wu L, Bohks SM, KewalRamani VN, Unutmaz D, Hope TJ (2003) Recruitment of HIV and its receptors to dendritic cell-T cell junctions. Science 300:1295–1297

McKimm-Breschkin J (2000) VP-14637 ViroPharma. Curr Opin Investig Drugs 1:425–427

MedImmune (2006) MedImmune (2006) Synagis package insert, MedImmune, Gaithersburg, MD. http://www.medimmune.com/products/bookmarks/synagis_pi.html. Cited 10 October 2006

Mimura K, Kono K, Hanawa M, Kanzaki M, Nakao A, Ooi A, Fujii H (2005) Trastuzumab-mediated antibody-dependent cellular cytotoxicity against esophageal squamous cell carcinoma. Clin Cancer Res 11:4898–4904

Monari C, Casadevall A, Pietrella D, Bistoni F, Vecchiarelli A (1999) Neutrophils from patients with advanced human immunodeficiency virus infection have impaired complement receptor function and preserved Fcgamma receptor function. J Infect Dis 180:1542–1549

Montefiori DC (1997) Role of complement and Fc receptors in the pathogenesis of HIV-1 infection. Springer Semin Immunopathol 18:371–390

Montefiori DC, Cornell RJ, Zhou JY, Zhou JT, Hirsch VM, Johnson PR (1994) Complement control proteins, CD46, CD55, and CD59, as common surface constituents of human and simian immunodeficiency viruses and possible targets for vaccine protection. Virology 205:82–92

Montefiori DC, Hill TS, Vo HT, Walker BD, Rosenberg ES (2001) Neutralizing antibodies associated with viremia control in a subset of individuals after treatment of acute human immunodeficiency virus type 1 infection. J Virol 75:10200–10207

Moore JP, Sattentau QJ, Klasse PJ, Burkly LC (1992) A monoclonal antibody to CD4 domain 2 blocks soluble CD4-induced conformational changes in the envelope glycoproteins of human immunodeficiency virus type 1 (HIV-1) and HIV-1 infection of CD4+ cells. J Virol 66:4784–4793

Mosley M, Smith-Burchnell C, Mori J, Lewis M, Stockdale M, Huan W, Whitcomb J, Petropoulos C, Perros M, Westby M (2006) Resistance to the CCR5 antagonist maraviroc is characterised by dose-response curves that display a reduction in maximal inhibition. 13th Conference on Retroviruses and Opportunistic Infections, Denver, CO

Murdoch C (2000) CXCR4: chemokine receptor extraordinaire. Immunol Rev 177:175–184

Murga JD, Franti M, Pevear DC, Maddon PJ, Olson WC (2006) Potent antiviral synergy between monoclonal antibody and small-molecule CCR5 inhibitors of human immunodeficiency virus type 1. Antimicrob Agents Chemother 50:3289–3296

Muster T, Steindl F, Purtscher M, Trkola A, Klima A, Himmler G, Ruker F, Katinger H (1993) A conserved neutralizing epitope on gp41 of human immunodeficiency virus type 1. J Virol 67:6642–6647

Nagashima KA, Thompson DA, Rosenfield SI, Maddon PJ, Dragic T, Olson WC (2001) Human immunodeficiency virus type 1 entry inhibitors PRO 542 and T-20 are potently synergistic in blocking virus-cell and cell-cell fusion. J Infect Dis 183:1121–1125

Nakowitsch S, Quendler H, Fekete H, Kunert R, Katinger H, Stiegler G (2005) HIV-1 mutants escaping neutralization by the human antibodies 2F5, 2G12, and 4E10: in vitro experiments versus clinical studies. Aids 19:1957–1966

Nishikawa M, Takashima K, Nishi T, Furuta RA, Kanzaki N, Yamamoto Y, Fujisawa J (2005) Analysis of binding sites for the new small-molecule CCR5 antagonist TAK-220 on human CCR5. Antimicrob Agents Chemother 49:4708–4715

Norris D, Morales J, Gathe J, Godofsky E, Garcia F, Hardwicke R, Lewis S (2006) Phase 2 efficacy and safety of the novel viral entry inhibitor, TNX-355, in combination with optimized background regimen (OBR). XVI International AIDS Conference, Toronto, Canada

Olson WC, Maddon PJ (2003) Resistance to HIV-1 entry inhibitors. Curr Drug Targets Infect Disord 3:283–294

Olson WC, Rabut GE, Nagashima KA, Tran DN, Anselma DJ, Monard SP, Segal JP, Thompson DA, Kajumo F, Guo Y, Moore JP, Maddon PJ, Dragic T (1999) Differential inhibition of human immunodeficiency virus type 1 fusion, gp120 binding, and CC-chemokine activity by monoclonal antibodies to CCR5. J Virol 73:4145–4155

Olson WC, Doshan H, Zhan C, Mezzatesta J, Assumma A, Czarnecky R, Stavola J, Maddon P, Kremer J, Israel R (2006) Prolonged coating of CCR5 lymphocytes by PRO 140, a humanized CCR5 monoclonal antibody for HIV-1 therapy. 13th Conference on Retroviruses and Opportunistic Infections, Denver, CO

Pantaleo G, Demarest JF, Vaccarezza M, Graziosi C, Bansal GP, Koenig S, Fauci AS (1995) Effect of anti-V3 antibodies on cell-free and cell-to-cell human immunodeficiency virus transmission. Eur J Immunol 25:226–231

Pantophlet R, Burton DR (2006) GP120: target for neutralizing HIV-1 antibodies. Annu Rev Immunol 24:739–769

Parren PW, Burton DR (2001a) The antiviral activity of antibodies in vitro and in vivo. Adv Immunol 77:195–262

Parren PW, Marx PA, Hessell AJ, Luckay A, Harouse J, Cheng-Mayer C, Moore JP, Burton DR (2001b) Antibody protects macaques against vaginal challenge with a pathogenic R5 simian/human immunodeficiency virus at serum levels giving complete neutralization in vitro. J Virol 75:8340–8347

Pastan I, Hassan R, Fitzgerald DJ, Kreitman RJ (2006) Immunotoxin therapy of cancer. Nat Rev Cancer 6:559–565

PDR (2006) The physicians' desk reference. http://www.pdr.net. Cited 10 October 2006

Perelson AS, Essunger P, Ho DD (1997) Dynamics of HIV-1 and CD4+ lymphocytes in vivo. Aids 11 [Suppl A]:S17–S24

Peterson A, Seed B (1988) Genetic analysis of monoclonal antibody and HIV binding sites on the human lymphocyte antigen CD4. Cell 54:65–72

Pierson TC, Doms RW (2003) HIV-1 entry and its inhibition. Curr Top Microbiol Immunol 281:1–27

Pincus SH (1996) Therapeutic potential of anti-HIV immunotoxins. Antiviral Res 33:1

Pincus SH, McClure J (1993) Soluble CD4 enhances the efficacy of immunotoxins directed against gp41 of the human immunodeficiency virus. Proc Natl Acad Sci USA 90:332–336

Pincus SH, Wehrly K, Tschachler E, Hayes SF, Buller RS, Reitz M (1990) Variants selected by treatment of human immunodeficiency virus-infected cells with an immunotoxin. J Exp Med 172:745–757

Pincus SH, Cole RL, Hersh EM, Lake D, Masuho Y, Durda PJ, McClure J (1991) In vitro efficacy of anti-HIV immunotoxins targeted by various antibodies to the envelope protein. J Immunol 146:4315–4324

Pincus SH, Cole R, Ireland R, McAtee F, Fujisawa R, Portis J (1995) Protective efficacy of nonneutralizing monoclonal antibodies in acute infection with murine leukemia virus. J Virol 69:7152–7158

Poignard P, Sabbe R, Picchio GR, Wang M, Gulizia RJ, Katinger H, Parren PW, Mosier DE, Burton DR (1999) Neutralizing antibodies have limited effects on the control of established HIV-1 infection in vivo. Immunity 10:431–438

Posner MR, Hideshima T, Cannon T, Mukherjee M, Mayer KH, Byrn RA (1991) An IgG human monoclonal antibody that reacts with HIV-1/GP120, inhibits virus binding to cells, and neutralizes infection. J Immunol 146:4325–4332

Pugach P, Kuhmann SE, Taylor J, Marozsan AJ, Snyder A, Ketas T, Wolinsky SM, Korber BT, Moore JP (2004) The prolonged culture of human immunodeficiency virus type 1 in primary lymphocytes increases its sensitivity to neutralization by soluble CD4. Virology 321:8–22

Purtscher M, Trkola A, Gruber G, Buchacher A, Predl R, Steindl F, Tauer C, Berger R, Barrett N, Jungbauer A, et al (1994) A broadly neutralizing human monoclonal antibody against gp41 of human immunodeficiency virus type 1. AIDS Res Hum Retroviruses 10:1651–1658

Ramachandran RV, Katzenstein DA, Wood R, Batts DH, Merigan TC (1994) Failure of short-term CD4-PE40 infusions to reduce virus load in human immunodeficiency virus-infected persons. J Infect Dis 170:1009–1013

Reichert JM, Rosensweig CJ, Faden LB, Dewitz MC (2005) Monoclonal antibody successes in the clinic. Nat Biotechnol 23:1073–1078

Reimann KA, Lin W, Bixler S, Browning B, Ehrenfels BN, Lucci J, Miatkowski K, Olson D, Parish TH, Rosa MD, Oleson FB, Hsu YM, Padlan EA, Letvin NL, Burkly LC (1997) A humanized form of a CD4-specific monoclonal antibody exhibits decreased antigenicity and prolonged plasma half-life in rhesus monkeys while retaining its unique biological and antiviral properties. AIDS Res Hum Retroviruses 13:933–943

Reimann KA, Khunkhun R, Lin W, Gordon W, Fung M (2002) A humanized, nondepleting anti-CD4 antibody that blocks virus entry inhibits virus replication in rhesus monkeys chronically infected with simian immunodeficiency virus. AIDS Res Hum Retroviruses 18:747–755

Robinson WE Jr, Montefiori DC, Mitchell WM (1988) Antibody-dependent enhancement of human immunodeficiency virus type 1 infection. Lancet 1:790–794

Roschke V, Clark S, Branco L, Kanakaraj P, Kaufman T, Yao X, Nardelli B, Shi Y, Cai W, Ullrich S, Bell A, Teng B, Lafleur DW, Chowdhury P, Kaithamana S, Sosnovtseva S, Albert V, Moore PA (2004) Characterization of a panel of novel human monoclonal antibodies that specifically antagonize CCR5 and block HIV-1 entry. 44th Annual Interscience Conference on Antimicrobial Agents and Chemotherapy, Washington, DC

Ruprecht RM, Hofmann-Lehmann R, Smith-Franklin BA, Rasmussen RA, Liska V, Vlasak J, Xu W, Baba TW, Chenine AL, Cavacini LA, Posner MR, Katinger H, Stiegler G, Bernacky BJ, Rizvi TA, Schmidt R, Hill LR, Keeling ME, Montefiori DC, McClure HM (2001) Protection of neonatal macaques against experimental SHIV infection by human neutralizing monoclonal antibodies. Transfus Clin Biol 8:350–358

Rusert P, Kuster H, Joos B, Misselwitz B, Gujer C, Leemann C, Fischer M, Stiegler G, Katinger H, Olson WC, Weber R, Aceto L, Gunthard HF, Trkola A (2005) Virus isolates during acute and chronic human immunodeficiency virus type 1 infection show distinct patterns of sensitivity to entry inhibitors. J Virol 79:8454–8469

Saifuddin M, Hedayati T, Atkinson JP, Holguin MH, Parker CJ, Spear GT (1997) Human immunodeficiency virus type 1 incorporates both glycosyl phosphatidylinositol-anchored CD55 and CD59 and integral membrane CD46 at levels that protect from complement-mediated destruction. J Gen Virol 78:1907–1911

Sanders RW, Venturi M, Schiffner L, Kalyanaraman R, Katinger H, Lloyd KO, Kwong PD, Moore JP (2002) The mannose-dependent epitope for neutralizing antibody 2G12 on human immunodeficiency virus type 1 glycoprotein gp120. J Virol 76:7293–7305

Sato H, Orenstein J, Dimitrov D, Martin M (1992) Cell-to-cell spread of HIV-1 occurs within minutes and may not involve the participation of virus particles. Virology 186:712–724

Sawyer LA (2000) Antibodies for the prevention and treatment of viral diseases. Antiviral Res 47:57

Scanlan CN, Pantophlet R, Wormald MR, Ollmann Saphire E, Stanfield R, Wilson IA, Katinger H, Dwek RA, Rudd PM, Burton DR (2002) The broadly neutralizing anti-human immunodeficiency virus type 1 antibody 2G12 recognizes a cluster of alpha1->2 mannose residues on the outer face of gp120. J Virol 76:7306–7321

Scanlan CN, Pantophlet R, Wormald MR, Saphire EO, Calarese D, Stanfield R, Wilson IA,
    Katinger H, Dwek RA, Burton DR, Rudd PM (2003) The carbohydrate epitope of the neutral-
    izing anti-HIV-1 antibody 2G12. Adv Exp Med Biol 535:205–218
Schrama D, Reisfeld RA, Becker JC (2006) Antibody targeted drugs as cancer therapeutics. Nat
    Rev Drug Discov 5:147–159
Shearer WT, Israel RJ, Starr S, Fletcher CV, Wara D, Rathore M, Church J, DeVille J, Fenton T,
    Graham B, Samson P, Staprans S, McNamara J, Moye J, Maddon PJ, Olson WC (2000)
    Recombinant CD4-IgG2 in human immunodeficiency virus type 1-infected children: phase
    1/2 study. The Pediatric AIDS Clinical Trials Group Protocol 351 Study Team. J Infect Dis
    182:1774–1779
Shearer WT, DeVille JG, Samson PM, Moye JH Jr, Fletcher CV, Church JA, Spiegel HM,
    Palumbo P, Fenton T, Smith ME, Graham B, Kraimer JM, Olson WC (2006) Susceptibility of
    pediatric HIV-1 isolates to recombinant CD4-IgG2 (PRO 542) and humanized mAb to the
    chemokine receptor CCR5 (PRO 140). J Allergy Clin Immunol 118:518–521
Shibata R, Igarashi T, Haigwood N, Buckler-White A, Ogert R, Ross W, Willey R, Cho MW,
    Martin MA (1999) Neutralizing antibody directed against the HIV-1 envelope glycoprotein
    can completely block HIV-1/SIV chimeric virus infections of macaque monkeys. Nat Med
    5:204–210
Stiegler G, Kunert R, Purtscher M, Wolbank S, Voglauer R, Steindl F, Katinger H (2001) A
    potent cross-clade neutralizing human monoclonal antibody against a novel epitope on
    gp41 of human immunodeficiency virus type 1. AIDS Res Hum Retroviruses 17:
    1757–1765
Stiegler G, Armbruster C, Vcelar B, Stoiber H, Kunert R, Michael NL, Jagodzinski LL, Ammann C,
    Jager W, Jacobson J, Vetter N, Katinger H (2002) Antiviral activity of the neutralizing anti-
    bodies 2F5 and 2G12 in asymptomatic HIV-1-infected humans: a phase I evaluation. Aids
    16:2019–2025
Stoiber H, Kacani L, Speth C, Wurzner R, Dierich MP (2001) The supportive role of complement
    in HIV pathogenesis. Immunol Rev 180:168–176
Stoiber H, Pruenster M, Ammann CG, Dierich MP (2005) Complement-opsonized HIV: the free
    rider on its way to infection. Mol Immunol 42:153–160
Strizki JM, Tremblay C, Xu S, Wojcik L, Wagner N, Gonsiorek W, Hipkin RW, Chou CC,
    Pugliese-Sivo C, Xiao Y, Tagat JR, Cox K, Priestley T, Sorota S, Huang W, Hirsch M, Reyes
    GR, Baroudy BM (2005) Discovery and characterization of vicriviroc (SCH 417690), a CCR5
    antagonist with potent activity against human immunodeficiency virus type 1. Antimicrob
    Agents Chemother 49:4911–4919
Sullivan BL, Knopoff EJ, Saifuddin M, Takefman DM, Saarloos MN, Sha BE, Spear GT (1996)
    Susceptibility of HIV-1 plasma virus to complement-mediated lysis. Evidence for a role in
    clearance of virus in vivo. J Immunol 157:1791–1798
Suntharalingam G, Perry MR, Ward S, Brett SJ, Castello-Cortes A, Brunner MD, Panoskaltsis N
    (2006) Cytokine storm in a phase 1 trial of the anti-CD28 monoclonal antibody TGN1412.
    N Engl J Med 355:1018–1028
Tagat JR, McCombie SW, Nazareno D, Labroli MA, Xiao Y, Steensma RW, Strizki JM, Baroudy
    BM, Cox K, Lachowicz J, Varty G, Watkins R (2004) Piperazine-based CCR5 antagonists as
    HIV-1 inhibitors. IV. Discovery of 1-[(4,6-dimethyl-5-pyrimidinyl)carbonyl]-4-[4-[2-
    methoxy-1(R)-4-(trifluoromethyl)phenyl]ethyl-3(S)-methyl-1-piperaz inyl]-4-methylpiperid-
    ine (Sch-417690/Sch-D), a potent, highly selective, and orally bioavailable CCR5 antagonist.
    J Med Chem 47:2405–2408
Takashima K, Miyake H, Kanzaki N, Tagawa Y, Wang X, Sugihara Y, Iizawa Y, Baba M (2005)
    Highly potent inhibition of human immunodeficiency virus type 1 replication by TAK-220, an
    orally bioavailable small-molecule CCR5 antagonist. Antimicrob Agents Chemother
    49:3474–3482
Takeda A, Tuazon CU, Ennis FA (1988) Antibody-enhanced infection by HIV-1 via Fc receptor-
    mediated entry. Science 242:580–583

Till MA, Zolla-Pazner S, Gorny MK, Patton JS, Uhr JW, Vitetta ES (1989) Human immunodefi-
    ciency virus-infected T cells and monocytes are killed by monoclonal human anti-gp41 anti-
    bodies coupled to ricin A chain. Proc Natl Acad Sci U S A 86:1987–1991
Trkola A, Pomales AB, Yuan H, Korber B, Maddon PJ, Allaway GP, Katinger H, Barbas CF, 3rd
    Burton DR, Ho DD, et al (1995) Cross-clade neutralization of primary isolates of human
    immunodeficiency virus type 1 by human monoclonal antibodies and tetrameric CD4-IgG.
    J Virol 69:6609–6617
Trkola A, Dragic T, Arthos J, Binley JM, Olson WC, Allaway GP, Cheng-Mayer C, Robinson J,
    Maddon PJ, Moore JP (1996a) CD4-dependent, antibody-sensitive interactions between HIV-
    1 and its co-receptor CCR-5. Nature 384:184–187
Trkola A, Purtscher M, Muster T, Ballaun C, Buchacher A, Sullivan N, Srinivasan K, Sodroski J,
    Moore JP, Katinger H (1996b) Human monoclonal antibody 2G12 defines a distinctive neu-
    tralization epitope on the gp120 glycoprotein of human immunodeficiency virus type 1. J Virol
    70:1100–1108
Trkola A, Ketas T, Kewalramani VN, Endorf F, Binley JM, Katinger H, Robinson J, Littman DR,
    Moore JP (1998) Neutralization sensitivity of human immunodeficiency virus type 1 primary
    isolates to antibodies and CD4-based reagents is independent of coreceptor usage. J Virol
    72:1876–1885
Trkola A, Ketas TJ, Nagashima KA, Zhao L, Cilliers T, Morris L, Moore JP, Maddon PJ, Olson
    WC (2001) Potent, broad-spectrum inhibition of human immunodeficiency virus type 1 by the
    CCR5 monoclonal antibody PRO 140. J Virol 75:579–588
Trkola A, Kuhmann SE, Strizki JM, Maxwell E, Ketas T, Morgan T, Pugach P, Xu S, Wojcik L,
    Tagat J, Palani A, Shapiro S, Clader JW, McCombie S, Reyes GR, Baroudy BM, Moore JP
    (2002) HIV-1 escape from a small molecule, CCR5-specific entry inhibitor does not involve
    CXCR4 use. Proc Natl Acad Sci U S A 99:395–400
Trkola A, Kuster H, Rusert P, Joos B, Fischer M, Leemann C, Manrique A, Huber M, Rehr M,
    Oxenius A, Weber R, Stiegler G, Vcelar B, Katinger H, Aceto L, Gunthard HF (2005) Delay
    of HIV-1 rebound after cessation of antiretroviral therapy through passive transfer of human
    neutralizing antibodies. Nat Med 11:615–622
Tsamis F, Gavrilov S, Kajumo F, Seibert C, Kuhmann S, Ketas T, Trkola A, Palani A, Clader JW,
    Tagat JR, McCombie S, Baroudy B, Moore JP, Sakmar TP, Dragic T (2003) Analysis of the
    mechanism by which the small-molecule CCR5 antagonists SCH-351125 and SCH-350581
    inhibit human immunodeficiency virus type 1 entry. J Virol 77:5201–5208
Umana P, Jean-Mairet J, Moudry R, Amstutz H, Bailey JE (1999) Engineered glycoforms of an
    antineuroblastoma IgG1 with optimized antibody-dependent cellular cytotoxic activity. Nat
    Biotechnol 17:176
UNAIDS (2006) Mother to child transmission. http://www.unaids.org/en/Issues/Affected_
    communities/mothertochild.asp. Cited 12 October 2006
van Meerten T, van Rijn RS, Hol S, Hagenbeek A, Ebeling SB (2006) Complement-induced cell
    death by rituximab depends on CD20 expression level and acts complementary to antibody-
    dependent cellular cytotoxicity. Clin Cancer Res 12:4027–4035
Veazey RS, Shattock RJ, Pope M, Kirijan JC, Jones J, Hu Q, Ketas T, Marx PA, Klasse PJ, Burton
    DR, Moore JP (2003) Prevention of virus transmission to macaque monkeys by a vaginally
    applied monoclonal antibody to HIV-1 gp120. Nat Med 9:343–346
Watson C, Jenkinson S, Kazmierski W, Kenakin T (2005) The CCR5 receptor-based mechanism
    of action of 873140, a potent allosteric noncompetitive HIV entry inhibitor. Mol Pharmacol
    67:1268–1282
Winston A, Boffito M (2005) The management of HIV-1 protease inhibitor pharmacokinetic
    interactions. J Antimicrob Chemother 56:1–5
Wolfe EJ, Cavacini LA, Samore MH, Posner MR, Kozial C, Spino C, Trapnell CB, Ketter N,
    Hammer S, Gambertoglio JG (1996) Pharmacokinetics of F105, a human monoclonal anti-
    body, in persons infected with human immunodeficiency virus type 1. Clin Pharmacol Ther
    59:662

Wong JK, Hezareh M, Gunthard HF, Havlir DV, Ignacio CC, Spina CA, Richman DD (1997) Recovery of replication-competent HIV despite prolonged suppression of plasma viremia. Science 278:1291–1295

Wu L, Gerard NP, Wyatt R, Choe H, Parolin C, Ruffing N, Borsetti A, Cardoso AA, Desjardin E, Newman W, Gerard C, Sodroski J (1996) CD4-induced interaction of primary HIV-1 gp120 glycoproteins with the chemokine receptor CCR-5. Nature 384:179–183

Wyatt R, Sodroski J (1998b) The HIV-1 envelope glycoproteins: fusogens, antigens, and immunogens. Science 280:1884–1888

Wyatt R, Kwong PD, Desjardins E, Sweet RW, Robinson J, Hendrickson WA, Sodroski JG (1998a) The antigenic structure of the HIV gp120 envelope glycoprotein. Nature 393:705–711

Xiang SH, Doka N, Choudhary RK, Sodroski J, Robinson JE (2002) Characterization of CD4-induced epitopes on the HIV type 1 gp120 envelope glycoprotein recognized by neutralizing human monoclonal antibodies. AIDS Res Hum Retroviruses 18:1207–1217

Zeitlin L, Cone RA, Moench TR, Whaley KJ (2000) Preventing infectious disease with passive immunization. Microb Infect 2:701

Zhang XQ, Sorensen M, Fung M, Schooley RT (2006) Synergistic in vitro antiretroviral activity of a humanized monoclonal anti-CD4 antibody (TNX-355) and enfuvirtide (T-20). Antimicrob Agents Chemother 50:2231–2233

Zhu P, Liu J, Bess J Jr, Chertova E, Lifson JD, Grise H, Ofek GA, Taylor KA, Roux KH (2006) Distribution and three-dimensional structure of AIDS virus envelope spikes. Nature 441:847–852

Zwick MB, Labrijn AF, Wang M, Spenlehauer C, Saphire EO, Binley JM, Moore JP, Stiegler G, Katinger H, Burton DR, Parren PW (2001) Broadly neutralizing antibodies targeted to the membrane-proximal external region of human immunodeficiency virus type 1 glycoprotein gp41. J Virol 75:10892–10905

Human Monoclonal Antibody and Vaccine Approaches to Prevent Human Rabies

T. Nagarajan, Charles E. Rupprecht, Scott K. Dessain,
P.N. Rangarajan, D. Thiagarajan, and V.A. Srinivasan(✉)

Abstract Rabies, being a major zoonotic disease, significantly impacts global public health. It is invariably fatal once clinical signs are apparent. The majority of human rabies deaths occur in developing countries. India alone reports more than 50% of the global rabies deaths. Although it is a vaccine-preventable disease, effective rabies prevention in humans with category III bites requires the combined administration of rabies immunoglobulin (RIG) and vaccine. Cell culture rabies vaccines have become widely available in developing countries, virtually replacing the inferior and unsafe nerve tissue vaccines. Limitations inherent to the conventional RIG of either equine or human origin have prompted scientists

V.A. Srinivasan
Indian Immunologicals Limited Gachibowli Post, Hyderabad, India
e-mail: srini@indimmune.com

S.K. Dessain (ed.) *Human Antibody Therapeutics for Viral Disease. Current Topics in Microbiology and Immunology 317.*
© Springer-Verlag Berlin Heidelberg 2008

to look for monoclonal antibody-based human RIG as an alternative. Fully human monoclonal antibodies have been found to be safer and equally efficacious than conventional RIG when tested in mice and hamsters. In this chapter, rabies epidemiology, reservoir control measures, post-exposure prophylaxis of human rabies, and combination therapy for rabies are discussed. Novel human monoclonal antibodies, their production, and the significance of plants as expression platforms are emphasized.

Abbreviations ABLV: Australian bat lyssavirus; BHK-21: Baby hamster kidney-21; CMV: Cytomegalovirus; CNS: Central nervous system; CVS: Challenge virus standard; DRV: DNA rabies vaccine; DUV: Duvenhage virus; EBLV: European bat lyssavirus; EBV: Epstein-Barr virus; ELISA: Enzyme-linked immunosorbent assay; ERIG: Equine rabies immunoglobulin; $F(ab')_2$: Divalent fragment antigen binding; Fab: Fragment antigen binding; Fc: Fragment crystallizable; GP: Glycoprotein; GT: Genotype; HDCV: Human diploid cell vaccine; HRIG: Human rabies immuno globulin; huMabs: Human monoclonal antibodies; IC-ELISA: Immunocapture ELISA   ID: Intradermal; IgG: Immunoglobulin G; LBV: Lagos bat virus; Mab: Monoclonal antibody; MNT: Mouse neutralization test; MOKV: Mokola virus; NTV: Nerve tissue vaccine; PAb: Polyclonal antibody; PCECV: Purified chick embryo cell vaccine; PEP: Post-exposure prophylaxis; PVRV: Purified vero cell rabies vaccine; RFFIT: Rapid fluorescent focus inhibition test; RIG: Rabies immunoglobulin; RRV: Rabies-related virus; RT-PCR: Reverse transcription polymerase chain reaction; RV: Rabies virus; scFv: Single chain variable fragment; SPR: Surface plasmon resonance; SRV: Street rabies virus; TSE: Transmissible spongiform encephalopathy

1   Introduction

Rabies continues to pose a serious public health threat worldwide, especially in developing countries. The vast majority of human fatalities reported globally come from India. Canine rabies continues to remain endemic in India, despite the availability of proven control measures. Rabies prevention is achieved either by pre- or post-exposure vaccination using modern, tissue culture-based vaccines. However, combined administration of both potent rabies vaccine and rabies immunoglobulin (RIG) is recommended in cases of all bites and mucosal exposures. The human RIG (HRIG) is widely used, especially in developed countries, and is considered safer than equine RIG (ERIG). The high cost of HRIG and its limited availability prohibit its wide use in developing countries such as India. The need to replace HRIG with at least an equally potent and a safer RIG product is considered to be important to improve the access to rabies biologicals. Fully human monoclonal antibodies (huMabs) with characteristics equivalent to that of

HRIG will provide an ideal alternative for prophylaxis of rabid animal exposures. When compared to various systems available, plants may offer the possibility of economical production of safe and efficacious RIG. Commercialization of economical RIGs in developing countries would result in their wider usage and reduced human rabies fatalities.

2   Clinical Manifestations of Rabies

Clinically, rabies presents a horrifying clinical picture and is essentially invariably fatal. About 80% of patients develop a classic or encephalitic (furious) form of rabies, and about 20% experience a paralytic (dumb) form of disease. About 50% to 80% of patients develop hydrophobia, which is a characteristic and widely recognized manifestation of rabies. Fever is common and may be quite high ($>107°F$), and there may be signs of autonomic dysfunction, including hypersalivation, lacrimation, sweating, piloerection, and dilated pupils (Hemachudha et al. 2002). Rabies mortality after untreated bites by rabid dogs varies from 38% to 57%, depending on the severity and location of the wound and presumably the virus concentration in the saliva (Sitthi-Amorn et al. 1987; Hemachudha 1989). Bites from rabid animals of other species, such as wolves, may result in 80% mortality or higher. The onset of infection depends on various factors such as the status of biting animal, site and number of bites, first aid received and time of initiation of rabies post-exposure prophylaxis (PEP) (Shetty et al. 2005). Bites on the head, face, neck, and hand, particularly with bleeding, carry the highest risk and are generally associated with a shorter incubation period. In the case of rabid bats, risk is present even after a superficial bite, owing partly to the unique ability of these viral agents to replicate in the epidermis and dermis (Morimoto et al. 1996).

The clinical course of rabies can be divided into several stages: incubation period, prodrome, acute neurological phase, coma, and death (Hemachudha 1989, 1994; Hemachudha and Phuapradit 1997). The typical incubation period of rabies is 1 to 3 months, but the range is from less than 7 days to more than 6 years (Smith et al. 1991; Hemachudha and Phuapradit 1997). The prodromal stage begins when the virus moves centripetally from the site of the bite to the dorsal root ganglia and to the central nervous system (CNS). These developments mark the end of the incubation period and symptoms at this stage are nonspecific and often include neuropathic pain at the bite site described as burning, numbness, tingling, or itching (Hemachudha and Phuapradit 1997; Hemachudha and Mitrabhakdi 2000). An intense and progressive local reaction, starting at the bite site and gradually spreading to involve the whole limb in a nonradicular pattern, or the ipsilateral side of the face, is a reliable indicator of rabies (Hemachudha 1989).

During the acute neurological phase, objective signs of nervous system dysfunction begin. Two-thirds of patients with classic rabies have an encephalitic form, and the remainder present with paralysis. Most patients with the encephalitic form die within 7 days (average 5 days) of onset, and the average survival period is about 2 weeks in paralytic cases (Hemachudha 1989; Hemachudha and Phuapradit 1997). The earliest feature of encephalitis is hyperactivity, aggravated by thirst, fear, light, noise, and other stimuli. Within 24 h, three major cardinal signs follow: fluctuating consciousness, phobic or inspiratory spasms, and autonomic stimulation signs. In contrast, paralytic rabies is characterized by persistent fever from the onset of limb weakness, intact sensory function of all modalities except at the bitten region, percussion myoedema, and bladder dysfunction (Hemachudha and Mitrabhakdi 2000). Coma precedes circulatory insufficiency (Kureishi et al. 1992). Attempted experimental treatments include interferon and antiviral drugs such as ribavirin, vidarabine, acyclovir, and inosine pranobex, as well as intrathecal and systemic high-dose administration of HRIG, steroids, and antithymocyte globulin; none has been successful (Hemachudha 1994). To date, most attempts to treat human rabies have been unsuccessful (Hemachudha et al. 2002), although a recent report described the survival of a young girl infected with rabies and treated with a combination of antiviral drugs and a drug-induced coma (Willoughby et al. 2005).

3   Rabies Virus

Rabies is caused by viruses in the genus *Lyssavirus*, within the family Rhabdoviridae. The *Lyssavirus* genus is subdivided into seven genotypes based on RNA sequencing (Bourhy et al. 1992, 1993; Amengual et al. 1997; Hooper et al. 1997; Badrane et al. 2001): genotype 1 (classical rabies virus, RV), genotype 2 (Lagos bat virus, LBV), genotype 3 (Mokola virus, MOKV), genotype 4 (Duvenhage virus, DUV), genotype 5 (European bat lyssavirus 1, EBLV-1), genotype 6 (European bat lyssavirus 2, EBLV-2), and genotype 7 (Australian bat lyssavirus, ABLV). Recently 4 new putative Lyssavirus genotypes have been described from Eurasian bats (Kuzmin et al. 2005). The members of genotypes 2–7 are known as rabies-related viruses (RRVs). All genotypes have caused human and/or animal deaths in nature (Badrane et al. 2001). The genotypes further segregate into two phylogroups: genotypes 1, 4, 5, 6, and 7 (phylogroup I); and 2 and 3 (phylogroup II) (Badrane and Tordo 2001). Viruses of each phylogroup differ in their biological properties, i.e., pathogenicity, induction of apoptosis, cell receptor recognition etc. (Badrane and Tordo 2001; Nadin-Davis et al. 2002). Rabies virus possesses a 12-kb single-stranded, nonsegmented RNA genome of negative polarity (Tordo et al. 1986). The viral genome encodes five structural proteins (3' to 5'): nucleoprotein (N), phosphoprotein (P), matrix protein (M), glycoprotein (G), and RNA-dependent RNA polymerase (L) (Wunner et al. 1988; Mebatsion et al. 1999). The 66 kDa glycoprotein coats the surface of the virus with approximately 400 trimeric spikes that are the major target for virus-neutralizing antibodies (Wunner 2002).

4 Rabies Epidemiology

4.1 Global

Rabies is considered to be a re-emerging zoonosis in many parts of the world (Taylor et al. 2001; Rupprecht et al. 2002), causing annually about 60,000 deaths worldwide (Martinez 2000), despite significant progress in vaccine development. Endemic rabies has been reported on every continent except Antarctica, but the vast majority of rabies deaths occur in Africa, Asia, and Latin America where animal control, vaccination programs, and effective human PEP are either unavailable or not effectively applied (Blancou 1988; Meslin et al. 1994; Kilic et al. 2006). In addition to mortality, rabies poses a major economic burden in developing countries as a result of the high cost of human PEP (Meslin et al. 1994) and loss of livestock (Ramanna et al. 1991; Nagarajan et al. 2006a). Dogs are the major reservoir of rabies in areas where canine rabies has not been controlled by vaccination (most of Asia, Africa, Latin America, and parts of the Caribbean) and are responsible for most of the human rabies deaths that occur worldwide (WHO 2002b; Kallel et al. 2006). The predominance of cases attributable to dogs reflects the high incidence of dog bites, estimated to be between 200 and 800 per 100,000 in many countries (Swaddiwuthipong et al. 1988). Approximately, 10 million people receive PEP after being exposed to rabies-suspect animals every year (Patrick and O'Rourke 1998; WHO 2006). Despite its global impact, rabies is not notifiable in most of the developing countries and is widely perceived as rare and unimportant by public health officials in many regions, especially Asia (Coleman et al. 2004; Ertl 2005). The World Health Organization ranks rabies low on its priority lists (WHO 2001).

Human rabies is uncommon in developed nations (WHO 1998), where systematic canine vaccination has led to a considerable decrease in the incidence of human exposures (Jenkins et al. 2004). Rabies virus continues to circulate mainly in wild animal hosts, presenting a more challenging problem (Fu 1997). In the United States, Canada, and most of Europe, infections in wild animals account for more than 95% of cases of rabies in animals, and the disease spills over to domestic animals and humans (Jenkins et al. 1988). In North America, reservoirs of rabies exist in many diverse animal species, including raccoons, skunks, bats, and foxes. In the United States a major rabies epizootic is associated with expansion of the raccoon rabies virus variants (Jenkins et al. 2004), and Canadian outbreaks are associated with the Arctic fox in the southern provinces of Canada, especially Ontario. The European rabies situation is largely associated with virus spread in the red fox, and the spread seems to be largely restricted to this species (Real et al. 2005). Gray foxes have been implicated as reservoirs of urban rabies posing direct threat to human health in parts of North and South America. Sylvatic rabies affecting mainly red foxes appeared during the mid-1940s at the Russian-Polish border and spread to other parts of Europe (Wandeler 2004). Rabies in domestic pets, such as dogs and cats, has been controlled in most parts of Europe, except in parts of the Russian Federation and Eastern Europe. Turkey is the only European country in which dog rabies is still prevalent and human rabies and rabies-suspect animal bites continue to pose public health problems (Kilic et al. 2006).

There is a growing incidence of human rabies cases transmitted by insectivorous bats in some parts of the world (i.e., the Americas and Australia), suggesting their importance in the epidemiology of rabies (Messenger et al. 2002; Mayen 2003; Kilic et al. 2006). Bats account for less than 20% of annual rabies cases reported in the USA and Canada, but during the last 20 years, variants of bat rabies have become the most common cause of human death from rabies (Hemachudha et al. 2002). Bat rabies has also been reported in several European countries, from Russia to Spain, particularly in coastal regions, and vampire bats are the major rabies reservoir in Latin America (Ito et al. 2003). Some countries, such as New Zealand, Japan, Taiwan, Finland, and Greece are free from rabies, as are Hawaii and many Pacific and Caribbean islands (excluding Cuba, the Dominican Republic, Haiti, Grenada, and Puerto Rico) (Chin 2000; WHO 2006). However, this freedom from rabies depends on the effective methods to prevent introduction of the virus and on active laboratory-based surveillance (Rupprecht and Gibbons 2004).

4.2   Asia

Rabies remains endemic in most of the developing countries of the Asian continent, including much of the Middle East, Pakistan, Afghanistan, India, Sri Lanka, Nepal, Bangladesh, Myanmar, Thailand, Laos, Cambodia, Vietnam, parts of Indonesia, the Philippines, and most of the former Soviet Republics. Greater than 95% of the approx. 50,000 cases are caused by dog bites, and more than 50% occur in children. Animals other than dogs can also pose a significant threat, as infected bats, mongooses, jackals, and other wildlife have been found in South and Southeast Asia (WHO 2004; Wilde et al. 2005). More than 3 cases of rabies per 100,000 population per year were reported in India and Sri Lanka in the 1970s (Bogel and Motschwiller 1986), and the Philippines reported 5–6 deaths per million population (Nishizono et al. 2002). Lack of animal rabies control programs, coupled with inadequate effective PEP, contributes to the high incidence of rabies in Asia (Knobel et al. 2005). Subsequent reports, however, have indicated that Asian countries such as the Philippines, Thailand, and Sri Lanka have been able to reduce rabies deaths recently to a great extent by application of effective PEP, but in India, Pakistan, and Bangladesh death rates have been stable (Sudarshan et al. 2007).

4.3   India

India has a prevalence of rabies that is among the highest in the world. In India, rabies is enzootic and widespread in all the states except in the island group of

Andaman, Nicobar, and Lakshadweep, which have consistently been free of rabies (Sudarshan et al. 2001). Based on community survey data, approximately 20,000 (or 2 per 100,000 population) human rabies deaths occur in India annually (WHO-APCRI 2004; Wilde et al. 2005). This is about 30% lower than an earlier estimate of 30,000 deaths (or 3 per 100,000 population) for the period 1992–2002, which was obtained from anecdotal data (WHO 2002b). Approximately 500,000 people undergo PEP in India, but only approximately 86% of bite cases receive rabies vaccination (Sudarshan et al. 2001).

Rabies in India occurs mainly in its urban form, in which stray dogs play an important role as a reservoir and transmitter of the disease to humans and domestic animals (Ramanna et al. 1991; Bhatia et al. 2004; Nagarajan et al. 2006a). Exposures may occur as a single event, or one rabid animal may expose multiple people or animals (Ramanna et al. 1991; Rotz et al. 1998). It is also evident from various reports that rabies maintenance and transmission happen in the sylvatic form, comprising various wildlife species, including foxes, wolves, mongooses, and others (Shah and Jaswal 1975; Bhatia et al. 2004; Matha and Salunke 2005). Overall, rabies in India is the result of contact with dogs in more than 96% of cases, jackals in 1.7%, cats in 0.8%, monkeys in 0.4%, mongooses in 0.4%, and foxes in 3% (Bhatia et al. 2004). Even rabid stray cattle have also been found to potentially transmit rabies to humans when ecological and societal factors are favorable (Nagarajan et al. 2006a). It is presumable that livestock animals and, rarely, human beings residing in villages along the forest areas are the common victims of wild animal bites and eventual contributors to the spread of rabies. Active surveillance studies would be able to unravel the actual existence of such a cycle (Kitala et al. 2000), but rabies is not a notifiable disease in India and there is no national program for animal rabies control and elimination.

To begin to assess rabies dynamics, vector ecology, transmission patterns, and the emergence of viral variants in India, we have begun molecular epidemiology studies (Nagarajan et al. 2006a). We obtained samples from different animal species (dog: 13 isolates; buffalo: 9 isolates; cow: 3 isolates; and goat: 4 isolates) and different geographical regions of India (Punjab, Andhra Pradesh, Karnataka, and Kerala). We used RT-PCR to amplify regions of the virus genome likely to contain meaningful genetic diversity: a 132-nucleotide region (G-CD), which encodes the cytoplasmic domain of the G gene, and a 549-nucleotide region (Psi-L), which includes sequences from the noncoding (Psi) region and the L gene. We found that the Indian RV isolates were genetically related to one another, with an average nucleotide similarity of greater than 95% in these regions. This observation was supported by similar data obtained by sequencing a domain of the N gene, in which all Indian RVs were identical to each other, thus defining a single major genetic cluster irrespective of the geographic region of origin (data not shown). Indian RV isolates could be distinguished from RV isolates from Asia, Europe, the Americas, and South Africa and from RRVs by the existence of three amino acids, 462G, 465H, and 468K. These three amino acids may be a useful epidemiological marker for RVs that have originated in the Indian subcontinent.

Within India, phylogenetic analyses of G-CD and Psi-L gene segments revealed that canine RV is the single major variant (Nagarajan et al. 2006a). The analyzed RV isolates form five genetic clusters ordered by geography, but not by host species. For instance, the RV isolates of GC4 and GC5 both were localized within Kerala, which is physically separated from the rest of India by the Western Ghats, but were found in three different hosts: goats, cattle, and dogs. Similar patterns of geographical restriction of RVs have been observed in southern Africa, Israel, and Middle Eastern countries (David et al. 2004; Nel et al. 2005). The predominance of canine RV circulation in India reflects the central role of the dog as the principal reservoir for rabies in India. Spread of canine rabies in India is likely facilitated by the high population density, constant supply of susceptible hosts, close proximity of donor species, inadequate herd immunity, and unrestricted animal movement (Holmes et al. 2002). These initial studies were limited to isolates from dogs and domestic animals. Future phylogenetic studies of RV isolates from human and wild animal cases will help to understand the nature of animal–human RV transmission and the possible role of the sylvatic cycle in the maintenance of endemic rabies. In addition, molecular phylogenetic studies will be essential to guide the creation and use of vaccine and antibody countermeasures that target the unique RV isolates circulating in India.

5   Reservoir Control and Vaccination

Globally, and particularly in developing countries, a combination of social, political, epidemiological, and technical factors have impeded the control of rabies. Much of the persistence of the disease results from the millions of unvaccinated stray or community-owned dogs that live in close association with humans (Wandeler et al. 1993). Movements of infected dogs to new uninfected areas have the potential to produce explosive, sustainable outbreaks (Childs et al. 2000) and may also result in spillover across geographical locations. Therefore, reduction of the prevalence of rabies in the canine reservoirs is the cornerstone of a multifaceted approach to control human rabies. Rabies, unlike any other viral zoonosis, offers the possibility of selective elimination, rather than mere control (Rupprecht et al. 2002). Epidemiological surveillance and the prevention of infection are essential to this process (Belotto et al. 2005). Efforts to monitor and control the incidence of rabies in its animal reservoirs will inform the development of optimized and affordable therapies for the prophylaxis and treatment of rabies exposure in humans. Strategies developed for the control of animal and human rabies in India can contribute to the process of canine rabies elimination in other developing countries.

Traditional approaches to dog rabies control have involved vaccination, movement restriction, and the culling of stray dogs (Cleaveland et al. 2003). In the past, various lethal techniques, including habitat destruction and extermination, have been attempted for culling stray dogs for rabies control. However, population reduction as the sole technique in disease abatement has been extremely difficult to justify in widespread, long-term, government-sponsored programs, and it raises serious economical, ethical, efficacy, and ecological issues (Rupprecht et al. 2002). Nonetheless, attempts have been made by municipal authorities in India to achieve rabies control through either culling or sterilization, but these efforts received a setback 4 years ago when animal rights activists influenced the Indian government to pass the Animal Birth Control Rules, which prohibit municipal authorities from killing stray animals (Mudur 2005).

A greater attention to prevention and control of animal rabies should lead to greater long-term benefits in public health than a narrow focus on human prophylaxis alone (Rupprecht et al. 2002). To control canine rabies in endemic areas, the WHO recommends mass vaccination of at least 80% of the dog population (WHO 2002a). Since a number of killed and live attenuated vaccines are now available for veterinary use that are safe, potent, and effective (Aubert et al. 1994; Fu 1997), at a small fraction of the cost of human vaccines, rabies prevention needs to remain focused clearly on disease control in the animal reservoir (Rupprecht et al. 2002). Mass vaccination of canines is the only way that could virtually eliminate rabies, but excessive bureaucracy, and technical and economical difficulties hamper this undertaking (Fu 1997). Nonetheless, the WHO reports indicate that approximately 50 million dogs are vaccinated annually against rabies throughout the world.

Delivery of an appropriate vaccine dosage can only be guaranteed by parenteral administration. In this case, current vaccines licensed for use in human and animals can confer protective immunity against a variety of street rabies virus (SRV) strains (Bourhy et al. 1992). This approach is not suitable for wild and stray animals. For mass vaccination, self-replicating virus vaccines are needed to contact the oral mucosa of a diversity of mammalian carnivores (Rupprecht et al. 2002). Oral immunization of stray dogs and wildlife against rabies using live vaccine is the most effective method to control and eventually eliminate rabies by induction of herd immunity. Identifying an accurate target for coverage is important because optimizing the cost-effectiveness of any vaccination program is likely to be a critical factor in determining the success and sustainability of control measures in countries with limited resources (Cleaveland et al. 2003). Spatial, temporal, and genetic heterogeneities that affect contact rate and transmission have important implications for the design of vaccination programs (May and Anderson 1984). During the last 30 years, great progress has been made in the development of oral vaccines against rabies. An ideal vaccine would protect against infection by all of the SRV strains that are associated with different mammalian species in diverse geographical locations (Morimoto et al. 2001a). More than optimal efficacy, safety is the most important criterion of any live vaccine. When used under field conditions, these vaccines must be safe for humans, the target species, and other animal species that may come in contact with the vaccine. Therefore, to be useful as a vaccine strain, pathogenicity of the virus must be highly attenuated without affecting the antigenic and immunogenic properties of the virus. This strategy has resulted in a substantial decrease of rabies enzooticity (Mitmoonpitak et al. 1998).

With the development of oral rabies vaccines for wildlife, control efforts can be focused on elimination of rabies among wildlife species serving as reservoirs (Velasco-Villa et al. 2005). Oral vaccination of free-ranging animals has been successfully used as an adjunct public health tool in both the developing and the developed world (Haddad et al. 1994; Perera et al. 2000; Estrada et al. 2001; Corn et al. 2003). Since the early 1980s, rabies vaccines, distributed by hand and by aircraft, have controlled rabies in foxes in parts of Western Europe, particularly Switzerland, Germany, Belgium, and France (Steck et al. 1982; Wandeler et al. 1988; Winkler and Bogel 1992). The primary focus of these efforts has been toward application in control against wildlife rabies in Europe and North America by strategic distribution of vaccine-laden baits (Sitthi-Amorn et al. 1987; Stohr and Meslin 1996; MacInnes et al. 2001). Conventional modified live vaccines, such as SAG-2 or poxvirus rabies glycoprotein recombinant vaccines, are very effective for oral immunization of foxes (Aubert et al. 1994), but they do not immunize skunks well or induce only low seroconversion by the oral route (Rupprecht et al. 1990). Very high doses of these vaccines may be necessary to induce protective immunity after oral immunization of dogs (WHO 1993), which makes an oral field vaccination impractical for economic reasons. Thus, although oral rabies virus vaccines have been used to immunize a diversity of mammalian carnivores, no single biological has yet been shown to be effective for all major species (Rupprecht et al. 2005).

Recently, advances in reverse genetics have allowed the design of recombinant RV (rRV) for consideration as new vaccines. These rRVs are as safe and effective as other commercial products in use for rabies prevention and control (Dietzschold et al. 1992; Dietzschold and Schnell 2002). To be useful as an oral vaccine, the recombinant virus must be able to replicate sufficiently not only after parenteral inoculation but, more importantly, after oral immunization. Safety, efficacy, and immunogenicity of recombinant RV vaccines were examined in captive dogs vaccinated by the oral route, compared to a commercial vaccinia rabies glycoprotein recombinant virus vaccine. Preliminary data demonstrated noninferiority of rRV products, suggesting that these vaccines hold promise for future development as oral immunogens for dogs and other important carnivore species (Rupprecht et al. 2002). The success of such control programs can be predicted according to an accurate identification of the species serving as a maintenance reservoir in the particular region targeted in a reservoir vaccination campaign. Accurate determination of patterns of disease transmission within and between different regions can suggest natural barriers to animal movement that can be exploited in a control program (Velasco-Villa et al. 2006).

6 DNA Vaccines

A variety of cell culture-derived vaccines are available for prophylaxis against rabies (Rupprecht et al. 2002). However, the perceived high cost of production may prohibit their wider use in developing countries. Clearly new concepts are needed to improve the potency, purity, safety, efficacy, stability, and economy of rabies vaccines. As such, a number of approaches are being explored for enhanced rabies prevention and control through molecular applications, including the creation of DNA vaccines (Lodmell and Ewalt 2001; Bahloul et al. 2006). The immunization of susceptible hosts with plasmid DNA is potentially a very promising way of conferring protection against infection (Hassett and Whitton 1996). The popularity of DNA vaccination is related in part to several characteristics, including relative ease of construction, elicitation of both humoral and cell-mediated immunity, adequate tolerance, versatility of delivery by multiple routes, and thermo-stability, which simplifies storage and shipment conditions (Wang et al. 1998). Furthermore, most bacterial plasmids provide their own adjuvant in the form of CpG motifs present in the bacterial DNA backbone (Pasquini et al. 1999).

Most research with DNA vaccines has focused upon the rabies virus G protein (RVGP), known to induce rabies virus neutralizing antibodies, which plays a critical role in immunity (Cox et al. 1977; Prehaud et al. 1988). To achieve high G protein expression levels, transcription of the RVGP gene is brought under the control of highly active promoters and enhancers, such as the cytomegalovirus (CMV) early promoter and enhancer (Bahloul et al. 1998) or the SV40 promoter (Tollis et al. 1991). Many groups have developed DNA vaccines encoding RVGP that provide protection against rabies in several experimental animal species (Tollis et al. 1991; Perrin et al. 1995; Xiang et al. 1995; Ray et al. 1997; Lodmell et al. 1998, 2001; Bahloul et al. 1998, 2006). Inoculation of mice with plasmids containing the gene encoding the RVGP efficiently induces humoral and cellular immune responses, resulting in protection against intracerebral challenge with the virus (Xiang et al. 1994). Full protection from a peripheral RV challenge has been observed in dogs 4 years after administration of a single dose of DNA rabies vaccine (DRV) (Bahloul et al. 2006). A DNA vaccine in newborn mice did not result in induction of tolerance but rather in induction of protective immunity (Wang et al. 1997). Immune responses in neonatal mice were only marginally affected by the presence of maternal immunity (Wang et al. 1998), suggesting that these vaccines may have utility in dogs less than 3 months old. Through recombinant DNA techniques, novel DNA vaccines can be readily created that may have improved potency and antiviral spectra. A DRV expressing both the ectodomain and the transmembrane domain (TD) of the RVGP may be an improved immunogen for generating high rabies virus neutralizing antibody (RVNA) titers (Rath et al. 2005). In addition, the spectrum of protection against lyssaviruses (including RV and RRV) has been broadened in mouse models by using chimeric lyssavirus glycoproteins (Bahloul et al. 1998; Jallet et al. 1999).

Historically, DRV has only demonstrated efficacy in pre-exposure rather than post-exposure situations. This could be due to the slow onset of antibody response following DNA vaccination (Lodmell et al. 1998). The usual response to DNA vaccination is a strong, durable, but slowly rising immune response to the encoded antigen (Lodmell et al. 1998). Several strategies are being examined to enhance the potency of DRV so that it can be used for both PEP and pre-exposure prophylaxis. The use of lymphokines such as granulocyte-macrophage colony-stimulating factor

(GM-CSF) and adjuvants such as monophosphoryl lipid A (MPL) has been shown to enhance the potency of DRV. A novel combination DNA vaccine containing a low dose of tissue culture-derived rabies vaccine and DRV has been tried and found to give complete protection against both peripheral and intracerebral RV challenge (Biswas et al. 2001). This co-inoculation approach seems to be a novel vaccination strategy for combating rabies in particular and infectious diseases in general (Biswas et al. 2001). Booster effects of DNA vaccines have been poor whether priming has been done by DNA vaccine or other vaccines (Ertl and Xiang 1996). However, restimulation with traditional vaccines has demonstrated the priming effects of a DNA vaccine (Ertl and Xiang 1996). Analyses of the pre- and post-exposure efficacy of DRV in mice (Ray et al. 1997; Bahloul et al. 1998; Wang et al. 1998), dogs (Perrin et al. 1999; Bahloul et al. 2006), and nonhuman primates (Lodmell et al. 1998) have yielded encouraging results, and several workers recently demonstrated post-exposure efficacy in a mouse model (Lodmell and Ewalt 2001; Bahloul et al. 2003). Since most rabies vaccinations in humans are initiated in post-exposure situations, it remains to be proved whether DNA vaccines are superior to currently used tissue culture rabies vaccines in human PEP (Dietzschold et al. 2003). Nonetheless, the WHO and FDA have expressed favorable recommendations toward the application of this technology to humans, so long as the necessary safety rules are applied (Center for Biologics Evaluation and Research 1995).

As a means of rabies control in animals, DNA-based vaccination may pave the way for effective national vaccination campaigns. Rather than revaccinating all dogs each year with tissue culture rabies vaccines as currently performed, which consumes vast quantities of human resources and is a costly task. The use of DNA vaccines may allow us to increase the time interval between two vaccine administrations, or even to rely on a single, lifelong injection and to target more specifically younger puppies and newly introduced dogs. The new vaccines may thus overcome the drawbacks of the established technologies (Bahloul et al. 2006). It would be optimal if strong immune responses to a DNA vaccine could be obtained rapidly after a single dose, because it is difficult to recover dogs for booster injections in developing countries (Bahloul et al. 2006). If potency can be sufficiently improved, then DNA vaccines may offer many advantages over classic technologies in terms of rapid development, simplicity, low cost, and broader immune responses. Accordingly, efforts are underway to enhance the immunogenicity of DNA vaccines by developing improved delivery systems or by combining DNA vaccines with adjuvants or tissue culture-derived vaccines (Jallet et al. 1999; Lodmell et al. 2000; Biswas et al. 2001; Lodmell and Ewalt 2001; Pinto et al. 2003). The fact that DNA vaccines can be produced more economically than the tissue culture-based vaccines and do not require a cold chain suggests that DNA vaccines may be ideally suited for canine rabies elimination programs in developing countries. In addition, the results of dog protection experiments demonstrate that DNA immunization is of great potential for protecting humans against RV by immunizing the canine reservoir against the disease (Perrin et al. 1999).

7 Human Rabies Vaccines

In countries where rabies is enzootic, most of the vaccines used are still produced from brain tissue, either from adult or suckling mammals (Yusibov et al. 2002), and are similar to the one developed by Louis Pasteur. The Semple rabies vaccine of sheep brain origin was developed by Sir David Semple in 1911 at the Central Research Institute, Kasauli, India. The presence of the myelin component in the Semple vaccine has resulted in severe neuroparalytic adverse reactions and even death in some recipients (Hemachudha et al. 1987; Meslin and Kaplan 1996). It holds an additional theoretical risk of transmission of transmissible spongiform encephalopathies (TSEs) from infected sheep to humans (Arya 1991). To overcome the neurologic complications of the Semple rabies vaccine, Fuenzalida and Palacios (1955) developed a rabies vaccine from brain tissue of newborn suckling mice, which is unmyelinated. The Fuenzalida rabies vaccine has been widely used in Latin American countries. These nerve tissue vaccines (NTVs) are primarily used in developing countries because they are less expensive than vaccines produced by tissue culture (Perrin et al. 1995). Unfortunately, the poor immunogenicity of NTVs increases the risk of vaccine failure (Perrin et al. 1999).

The first highly successful modern cell-culture vaccine was produced in the 1960s in human diploid cells by Wiktor et al. (1969). The human diploid cell vaccine (HDCV) produced significantly higher immunogenicity and less allergic reactions compared with first generation NTVs (Wiktor et al. 1969; Plotkin 1980). Clearly HDCV is highly effective in protection against rabies and rabies-related lyssavirus strains (Brookes et al. 2005). Although the HDCV produces high serologic titers, lower virus yields and higher production costs make this vaccine unaffordable in developing countries of the world, where the majority of human deaths occur (Plotkin et al. 1999). Economical and technical hurdles have hampered the use of these tissue culture rabies vaccines in developing countries (Trabelsi et al. 2006). In the 1980s a second cell culture vaccine, primary chick embryo cell vaccine (PCECV), was developed (Barth et al. 1984). The PCECV is highly purified and is as effective as HDCV (Dreesen et al. 1989). It is used worldwide in both industrialized and developing countries. Another of the most reliable and economical human rabies vaccine is purified Vero cell rabies vaccine (PVRV) produced from Vero, a continuous cell line. The Vero cell line has a long and successful history of being used in the production of rabies and polio vaccines worldwide (Montagnon et al. 1983). The highly purified PVRV is manufactured in France, Columbia, China, and India and is widely used in Europe and Asia. The virus titer produced by Vero cells is higher than the titer produced by human diploid cells and hence the final product is less expensive. Large-scale propagation using improved tissue culture technology (e.g., microcarriers and bioreactors) will result in further reduction in the cost of rabies vaccines (Dietzschold et al. 2003).

Currently in India, six modern cell culture vaccines are available, namely HDCV (Rabivax, Serum Institute of India, Pune), PCECV (Rabipur, Chiron Behring, Ankleshwar), PVRV (Abhayrab, Human Biologicals Institute,

Udhagamandalam; Rabirix, Bharat Biotech International Limited, Hyderabad), and Verorab, (Aventis Pasteur, Lyon), and DEV (Vaxirab, Zydus Cadila, Ahmedabad). Abhayrab has recently been shown to be a highly immunogenic and a safe vaccine (Sampath et al. 2005). The production methods of both the human and veterinary rabies vaccines remain the same with a few minor variations in upstream and downstream processes (Table 1). Rabies vaccine production requires stringent quality control to ensure adequate safety and potency. Viral inactivation is a critical step in the manufacture of rabies vaccines, since the presence of uninactivated virus in the final vaccines can cause post-vaccinal accidents. There are instances in which rabies vaccines had to be recalled after live rabies virus contamination was suspected in a commercial vaccine (Centers for Disease Control and Prevention 2004). This stresses the significance of optimizing inactivation conditions and studying the inactivation kinetics. We have validated the inactivation process by studying the inactivation kinetics for beta propiolactone (BPL) used in the manufacture of human rabies vaccines by testing the antigen at different intervals after addition of BPL for infectivity titers (M. Elaiyaraja, T. Nagarajan, G.S. Reddy, D. Thiagarajan, and V.A. Srinivasan, unpublished data). One of the tests gaining significance is estimation of RVGP content of rabies vaccines as an in-process control in addition to the NIH potency test. Measurement of RVGP content also helps to maximize the

Table 1 Comparison of tissue culture-based human and veterinary rabies vaccines

S. No.	Item	Human rabies vaccine	Veterinary rabies vaccine
1	Vaccine virus strain	PV/PM/Flury LEP	CVS/Flury LEP/SAD/PM
2	Cell substrate	Vero/CEF/MRC-5/Duck embryo	BHK-21/PHK
3	Cell type	Anchorage dependent	Anchorage independent/ suspension
4	Production system	Roller culture/microcarrier culture	Suspension culture— bioreactor
5	Inactivant	BPL	BEI/BPL
6	Vaccine type	Freeze dried/liquid	Liquid
7	Adjuvant	No adjuvant in freeze dried vaccines; liquid vaccines contain aluminum hydroxide gel	Aluminum hydroxide gel
8	Purification technique	Rate zonal ultracentrifugation/ chromatography	No elaborate downstream processing
9	Potency	2.5 IU/dose	1 IU/dose
10	Duration of immunity	1 year	3 years for nonendemic countries and 1 year for endemic countries
11	Cost	Expensive	Inexpensive
12	Special requirements	Freedom from host cell DNA and BSA	None

BEI, binary ethyleneimine; BPL, beta propiolactone; BSA, bovine serum albumin; CEF, chick embryo fibroblast; S. No., serial number

recovery of antigen during vaccine production, which may help to reduce the cost of production. To this end, we have developed an immunocapture ELISA (IC-ELISA) for GP measurement that uses a novel murine Mab specific for the RVGP (Nagarajan et al. 2006b). RVGP content of our Abhayrab vaccine, as measured with the IC-ELISA, correlates closely with established NIH potency estimates (Nagarajan et al. 2006b).

Rabies vaccines produced by tissue culture techniques are safe and effective in humans for PEP and pre-exposure prophylaxis and have significantly reduced the worldwide incidence of human rabies. The economic and human benefits of vaccination are well documented (Ulmer and Liu 2002). Commercial rabies vaccines for human use are compo sed of inactivated RV strains and are fully protective against RV (>99%) (Koprowski et al. 1985; Bourhy et al. 1992). The RV immune response also elicits cross-protection against the lyssavirus genotypes, principally those in phylogroup I. Rabies vaccines have not been shown to provide cross-protection for other lyssaviruses in genotype 2 and 3, LBV and MOKV, respectively, which are those in phylogroup II (Badrane et al. 2001). Tissue culture rabies vaccines are highly purified and therefore produce few severe allergic reactions (Nicholson 1996). Significantly, they also induce less pain, require fewer doses, and produce fewer side effects than the lower potency NTVs (Hemachudha et al. 1987; Swaddiwuthipong et al. 1988; Parviz et al. 1998). In some developing countries, NTVs have already been replaced through the use of tissue culture rabies vaccine following reduced dosage intradermal (ID) regimens for PET. The ID regimens have proved to be efficacious and cost-effective replacements for NTV (Chutivongse et al. 1990; Briggs et al. 2000). Currently, almost 60% of rabies-exposed people take one of the modern tissue culture vaccines (Ichhpujani et al. 2001) and nearly 5 million doses of these vaccines are sold every year (Sudarshan et al. 2007). Semple vaccine production in India was suspended in December 2005.

While vaccines from tissue culture are undoubtedly the vaccines of choice, it must be recognized that they are currently inaccessible to a significant part of the developing world. Despite the presence of at least six brands of modern cell culture vaccines in the Indian market, the inequalities in distribution and the suboptimal use of these vaccines are striking. For economic reasons, varying ID regimens are also used for PEP (Warrell et al. 1985; Suntharasamai et al. 1994; Briggs et al. 2000). In India, most of the patients undergoing PEP are given only 3 doses instead of the 5 doses of the Essen schedule. Though people perceive rabies as a dreaded disease, most dog bites result in an initial visit to the physician for advice and primary vaccination. However, the patients may fail to complete the series of vaccinations if the animal appears to remain healthy after the biting incident. The administration of RIG will depend on the earlier history of vaccination of individuals. Patients who have received three doses, which can be considered a prophylactic vaccination, may not need to be administered RIG along with rabies vaccine in cases of future rabies exposures. The consequence of not properly vaccinating against rabies is often not fully understood by the public. The future strategy for human rabies vaccination should therefore aim to educate the public as well as to ensure a stable, affordable, and universally accessible supply of existing vaccines.

Our approach to these problems has been to establish a series of Abhay clinics to provide the Abhayrab vaccine. These franchise clinics are operated by qualified medical practitioners who provide disease education while ensuring that the cold chain is maintained and that the vaccine is properly administered.

8   Post-exposure Prophylaxis

Rabies is one of the few infectious diseases that can be prevented by PEP. According to the WHO guidelines, category 3 exposures to rabies, which are defined as either single or multiple transdermal bites or contamination of mucous membranes with saliva of a rabid animal, require rabies PEP (WHO 2006). PEP includes immediate local treatment of the wound with washing and disinfection, local wound infiltration with therapeutic RIGs, and vaccination. The efficacy of PEP is probably due to the long incubation period of the disease (Lafon et al. 1990). The principal modality of protection from rabies exposure by vaccination is the development of neutralizing antibodies, which are primarily directed against the viral surface G protein (Wunner 2002). Induction of these antibodies by the vaccine is a key determinant of viral neutralization and animal protection against disease development, and must occur before the RV has entered the relatively immunoprotected CNS (Cox et al. 1977; Baer and Lentz 1991; Cliquet et al. 1998). The measurement of neutralizing antibodies to RV is commonly used to assess the level of immunity to rabies in animals and humans. The appropriate administration of modern rabies biologics according to approved regimens virtually assures protection against the development of rabies (Rotz et al. 1998). The success of PEP in humans is highly dependent on the punctuality of intervention, the severity and location of the wounds, the quality of the vaccine, and proper local instillation of RIG into all the wounds in a timely manner after exposure (Servat et al. 2003). Most human rabies cases occur because of one or more of the following reasons: (1) no PEP of any kind is used; (2) RIG is unavailable or unaffordable; (3) prophylaxis is significantly delayed or inappropriate; or (4) acute illness, malnutrition, or other underlying conditions compromise the immune response (Hanlon et al. 2001).

The life-saving benefit of combining rabies immune serum with vaccine was first shown by Balthazard et al. (1955), and has been validated in many follow-up clinical studies, particularly for category 3 exposures (Dean et al. 1963; Chutivongse et al. 1991; Strady et al. 1996; Wilde et al. 1996). For transdermal or mucosal exposure to rabies virus, it is recommended that as much as possible of the RIG be instilled into the bite site (WHO 1996). Although the mechanism by which RIG neutralizes RV remains unclear, the local infiltration of virus inoculation sites has been shown to be a key element to confer protection in humans, particularly in previously unimmunized subjects (Khawplod et al. 1996; Servat et al. 2003). It is likely that the combined effects of RVNA and an active vaccine-induced immune response work together to clear the virus before access to CNS can occur (Wilde et al. 1989; Wilde et al. 1996).

Laboratory studies have demonstrated that use of vaccine alone does not prevent lethal RV infection in severe post-exposure situations (Koprowski et al. 1950; Sikes et al. 1971), and study of attenuated RV infections in immunocompromised mice has shown that cellular and humoral mechanisms collaborate in viral clearance (Hooper et al. 1998). This perhaps explains the need for both passive administration of RIG and vaccination for reliable PEP of humans, despite the fact that under some conditions passive immunization may possibly interfere with the induction of an active immune response to RV (Hattwick et a l. 1976; Rupprecht et al. 1992).

9   Conventional Polyclonal Rabies Immune Globulins

Either polyclonal ERIG or HRIG is recommended for use in PEP (Steele 1988). Historically, hyperimmune polyclonal antibodies (PAb) prepared from pooled sera have been used extensively, largely due to the multitude of antibody specificities generated in the typical polyclonal response. PAb are generally relatively easy to make, provided the immune donors are available. The PAb preparations have the advantage of including antibodies of various specificities and isotype that provide diversity in biological function through various constant regions. Due to their polyvalent characteristics and superior ability to eliminate or neutralize complex target antigens, blood-derived IgGs remain the preferred therapeutic choice for many conditions (Haurum and Bregenholt 2005). The therapeutic implication is that PAb reacting with several epitopes of the same viral protein may be superior to a Mab that inherently only reacts with a single viral epitope (Casadevall 2002; Brekke and Sandlie 2003).

HRIG has a safety profile superior to that of ERIG and is ideally used in all cases of PEP (McKay and Wallis 2005). Unfortunately, more than 90% of rabies cases are encountered in developing countries where HRIG is not readily affordable or available and ERIG is the only option (Meslin et al. 1994; Wilde et al. 1996; McKay and Wallis 2005). Potential complications of ERIG administration include serum sickness (bone pain, renal failure, and encephalopathy), severe allergic reactions, and life-threatening anaphylaxis. Many modern ERIG formulations consist of purified, heat-treated, $F(ab')_2$ products that have a lower rate of adverse effects than crude serum, although there is a theoretical weakness of $F(ab')_2$ products in severe exposures due to their rapid clearance and the lack of the Fc region. In addition to the safety concerns associated with ERIG, some animal ethics groups oppose the rearing of animals for serum production (Wilde et al. 2002).

Despite their intrinsic advantages, HRIG products are expensive to manufacture and have several limitations, such as variable pharmacological profiles (Bregenholt and Haurum 2004), lot-to-lot variation in the amount of rabies virus-specific antibody, the possibility of transmission of infectious agents (Slade 1994), and a chronically limited worldwide supply (Farrugia and Poulis 2001). The limitation in supply is related in part to inherent imprecision in the capacity to predict future demands, as well as limited recruitment of new donors and a substantial production delay from the bleeding of donors to release of the final product (Wilde et al. 1996). We have considered the possibility of creating a similar HRIG product indigenously to increase its availability and lower its cost. However, this could be expensive and technically complex, due to problems associated with recruitment of healthy donors, the high cost involved in the screening of donors and batches of sera for identified pathogens, and the possibility of transmission of unidentified blood-borne pathogens. These difficulties encouraged us to embark on the development of human monoclonal antibodies (huMabs) against RV that can be used as an alternative to ERIG and HRIG.

10   Human Monoclonal Antibodies for Rabies PEP

10.1   Combination Therapy for Rabies PEP

An ideal antibody preparation for rabies PEP intended for use in India and other developing countries must meet a number of essential criteria. First, it must be able to neutralize the viruses likely to be encountered in the geographical area in which it is to be used. Ideally, clinical use of a cloned RIG would be combined with ongoing epidemiological surveillance of RV and RRV strains (Krebs et al. 2005; Nagarajan et al. 2006a). Second, it should contain at least two antibodies that bind nonoverlapping and noncompeting epitopes, so that mutant viruses able to escape the neutralization of one antibody will still be neutralized by the other (Marissen et al. 2005; Goudsmit et al. 2006). Third, the antibodies must be nonimmunogenic, to reduce the risk of first-dose side effects and to enable repeat dosing, if necessary. For this purpose, fully human antibodies are required. Fourth, it must have efficacy comparable to existing HRIG preparations in animal models of RV infection (Goudsmit et al. 2006; de Kruif et al. 2007). Finally, it must be produced at a cost that makes it affordable for the individuals who require it, many of whom are impoverished yet are required to pay for PEP out-of-pocket.

The ability of cloned antibodies to neutralize RVs is well-established with both murine (Wiktor and Koprowski 1978; Libeau and Lafon 1984; Schumacher et al. 1989; Nagarajan et al. 2006b) and also human antibodies (Lafon et al. 1990; Ueki et al. 1990; Enssle et al. 1991; Champion et al. 2000; Kramer et al. 2005). Neutralizing antibodies bind to one of three antigenic regions of the G protein, which forms the outer surface of the virion (Benmansour et al. 1991; Marissen et al. 2005). The mechanisms of antibody-mediated neutralization of the virus are incompletely understood, but appear to involve complex mechanisms that both inhibit viral entry into its target neurons as well as the ability of the virus to function once it has entered the cell, perhaps by inhibiting the exit of the virus from the endosomal vesicles (Dietzschold et al. 1987, 1992). Short single-chain (scFv) and dimeric, antigen-binding domain-only (Fab) antibodies are able to neutralize virus in vitro, indicating a capacity to directly interfere with viral function (Cheung et al. 1992; Muller et al. 1997; Kramer et al. 2005). For efficient rabies virus clearance from the CNS, full-length antibodies collaborate with cell-mediated inflammatory processes (Hooper et al. 1998).

Human antibodies capable of neutralizing the RV in vivo have been isolated by recombinant DNA (phage display) and hybridoma methods (Wiktor et al. 1969; Dietzschold et al. 1990; Ueki et al. 1990; Enssle et al. 1991; Champion et al. 2000; Kramer et al. 2005). The first neutralizing antibodies were cloned by creating hybridomas between Epstein-Barr virus (EBV)-immortalized primary human B cells, obtained from rabies-immune volunteers, and murine myeloma or heteromyeloma cell lines (Dietzschold et al. 1990; Lafon et al. 1990; Ueki et al. 1990; Enssle et al. 1991; Champion et al. 2000; Kramer et al. 2005). Prosniak et al. (2003) first combined three cloned human antibodies to create a "cocktail" capable of comprehensively neutralizing a variety of fixed and street RVs. The cocktail effectively provided PEP in rabies-exposed mice and had an efficacy comparable to HRIG (Prosniak et al. 2003). The two most potent of these antibodies (CR57 and CRJB) were studied further. Unfortunately, the two antibodies were found to bind overlapping epitopes on antigenic domain I of the RVGP, such that mutant virus strains that had acquired resistance to one antibody were often found to be resistant to the other (Marissen et al. 2005). A potentially cross-protective antibody was sought through a phage display cloning experiment that used RV-immune human peripheral blood lymphocytes as starting material, producing 21 novel human IgG antibodies capable of RV neutralization in vitro (Kramer et al. 2005). One of these antibodies, CR 4098, was found to interact with antigenic domain III of the RVGP and to have optimal association and dissociation kinetics by surface plasmon resonance (SPR) analysis. Cross-protection with the two antibodies was demonstrated, such that the second antibody neutralized mutant viruses escaping neutralization by the other antibody. Thus, these two antibodies fulfilled two essential requirements of an effective huMab cocktail, including broad strain reactivity and the ability to bind nonoverlapping, cross-protective G protein epitopes (Kramer et al. 2005). Accordingly, the antibody combination was found to be effective in a Syrian hamster PEP model and to be of comparable potency to standard HRIG. Importantly, the combination did not interfere with the development of immunity induced by concomitantly administered vaccine (Goudsmit et al. 2006; de Kruif et al. 2007).

10.2   Novel Human Monoclonal Antibodies Suitable for Use in the Indian Subcontinent

We sought to build on our murine monoclonal antibody experience to develop novel human antibodies suitable for use in RIG tailored to the RV isolates circulating in the Indian subcontinent. For this purpose, we created heterohybridomas between primary peripheral blood B cells and a heteromyeloma cell line, K6H6/B5 (Carroll et al. 1986). We chose the K6H6/B5 cell line because it has been successfully used to clone human antiviral antibodies (Siemoneit et al. 1994; Funaro et al. 1999). We used mitogen stimulation, rather than EBV transformation, to prepare the primary B cells for fusion to reduce the likelihood that the heterohybrid cell lines would express EBV antigens that may contaminate antibodies purified for human use.

We obtained peripheral blood lymphocytes from a subject multiple vaccinated with the Abhayrab vaccine, which incorporates the PV strain and is known to provide broad coverage against RV strains (Badrane et al. 2001).

Using the hybridoma method, we have cloned eight new human IgG antibodies (huMabs) that bind the RVGP. We first characterized the isotype, specificity, and crossreactivity of each of the huMabs. Each of the huMabs was of gamma 1 (γ1) heavy chain and lambda (λ) light chain isotype. Their specificity to RV was demonstrated by a cell-ELISA using unfixed mock and RV-infected cultured cells. All the huMabs showed reactivity to all the fixed RVs except the BHK-21-adapted CVS strain. None of the huMabs showed any reactivity to host cell protein (see Table 2). The huMabs bound specifically to the native form of whole virus antigens as well as purified RVGP and did not react with ribonucleoprotein (RNP) as evident from the results of indirect ELISA (see Table 3). The huMabs recognized

Table 2 Reactivity of anti-rabies virus human monoclonal antibodies with various fixed RV strains in cell-ELISA[a]

S. No.	HuMab	PV	CVS-11	SAD	Flury LEP	CVS-BHK	Host cell
		Fixed RV strains					
1	R17D6	+	+	+	+	-	-
2	R14D3	+	+	+	+	-	-
3	R16C9	+	+	+	+	-	-
4	R14D6	+	+	+	+	-	-
5	R18G9	+	+	+	+	-	-
6	R16F7	+	+	+	+	-	-
7	R17G9	+	+	+	+	-	-
8	R16E5	+	+	+	+	-	-

[a]Unfixed rabies virus infected and mock-infected cell culture was used as the solid phase antigen for doing cell-ELISA to demonstrate the specificity of huMabs to rabies virus

Table 3 Characteristics of anti-rabies virus human Mabs

		Isotype		Native (strain PV)			Denatured (strain PV)			Antigenic site
S. No.	HuMab	Heavy chain	Light chain	Whole virus	GP	RNP	Whole virus	GP	RNP	
1	R17D6	γ1	λ	+	+	-	-	-	-	III
2	R14D3	γ1	λ	+	+	-	-	-	-	III
3	R16C9	γ1	λ	+	+	-	-	-	-	III
4	R14D6	γ1	λ	+	+	-	-	-	-	III
5	R18G9	γ1	λ	+	+	-	-	-	-	III
6	R16F7	γ1	λ	+	+	-	-	-	-	III
7	R17G9	γ1	λ	+	+	-	-	-	-	III
8	R16E5	γ1	λ	+	+	-	-	-	-	III

antigenic site III of RVGP as determined by a competitive ELISA using a mouse Mab (D1) against antigenic site III as described elsewhere (Nagarajan et al. 2006b).

We tested the ability of our huMabs to neutralize various street and fixed RVs both in vitro and in vivo. In the mouse neutralization test (MNT) with Swiss albino mice, the huMabs neutralized all the four Indian SRVs of dog origin (108, 129, 141, and 142) (Table 4). All the huMabs neutralized four fixed RV strains (PV, Flury LEP, SAD, and CVS-11) while none of the huMabs neutralized BHK-21-adapted CVS RV when tested by the rapid fluorescent focus inhibition test (RFFIT). The huMab R17D6 had the highest RFFIT titers of our tested antibodies, ranging from 2.5 IU to 13.5 IU. The titers of the other huMabs were lower (Table 5). We tested the ability of the five huMabs with the most potent RFFIT titers to neutralize members of other *Lyssavirus* genotypes. All four huMabs neutralized GT7 (*ABLV*) and GT4 (*DUV*) effectively in the RFFIT, although none neutralized GT3 (*MOKV*) or GT5 (*EBLV-1*) (Table 6).

Dietzschold and colleagues have repeatedly emphasized the importance of evaluating the suitability of antibodies for use in rabies prophylaxis by testing in animal models (Dietzschold et al. 1990, 1992). We tested whether two of our antibodies, alone and in combination, could enhance the efficacy of the Abhayrab vaccine in conferring PEP to Syrian hamsters inoculated with the Indian SRV 129 and

Table 4 Survivorship of Swiss albino mice subjected to mouse neutralization test[b] to demonstrate the ability of human monoclonal antibodies to neutralize various fixed and street rabies viruses in vivo

		Virus			
S. No.	huMab[a]	SRV 108	SRV 129	SRV 141	SRV 142
1	R17D6	100%	100%	100%	100%
2	R14D3	ND	100%	ND	ND
3	R16C9	100%	100%	100%	100%
4	R14D6	100%	100%	100%	100%
5	R18G9	100%	100%	100%	100%
6	R16F7	100%	100%	100%	100%
7	R17G9	100%	100%	100%	100%
8	R16E5	100%	100%	100%	100%

*Not done

[a] Heat inactivated heterohybridoma culture supernatant containing huMabs was used for neutralization reaction

[b] Female Swiss albino mice (3–4 weeks old) were intracranially inoculated with 30 µl of (50 MICLD$_{50}$/30 µl) either rabies virus infected culture fluid or brain homogenates from naturally infected rabid stray dogs. The inoculated mice were observed daily for symptoms typical of rabies for 21 days. The results are expressed as percentage of mice that survived after 21 days of observation

Table 5 Neutralization of rabies viruses by anti-rabies virus human monoclonal antibodies

S. No.	huMab[b]	PV	CVS-11	SAD	Flury LEP	CVS-BHK
		VNA[a] (IU/ml)				
		Fixed RV strains				
1	R17D6	2.6	11.5	13.8	10.0	-
2	R14D3	2.6	2.8	4.8	3.1	-
3	R16C9	2.8	3.5	12.9	0.3	-
4	R14D6	0.8	1.6	2.8	1.3	-
5	R18G9	2.8	5.2	12.9	8.1	-
6	R16F7	2.6	2.8	6.3	4.8	-
7	R17G9	2.6	2.0	2.8	3.0	-
8	R16E5	0.7	2.0	2.8	2.6	-

[a] VNA titer was determined by RFFIT using rabies virus strain CVS-11 essentially as described by Smith et al. (1996)
[b] Heterohybridoma culture supernatant containing huMabs was used after heat inactivation. None of the huMabs tested could neutralize the rabies virus strain CVS-BHK

Table 6 Neutralization of rabies-related viruses by anti-rabies virus human monoclonal antibodies

S. No.	huMab[b]/SRIG	Mokola (GT3)	DUV (GT4)	EBV1 (GT5)	ABV (GT7)
		VNA[a] (IU/ml)			
		Rabies-related viruses			
1	R16C9	<5	7	<5	250
2	R16E5	<5	7	<5	250
3	R16F7	<5	6	<5	250
4	R14D6	<5	<5	<5	145
5	SRIG	<5	9	11	56

[a] VNA titer was determined by RFFIT at CDC, Atlanta, essentially as described by Smith et al. (1996)
[b] Semipurified huMAb preparation was used

SRV 141 strains. Both the R16F7 and R14D6 huMabs provided 100% protection against both of these strains, whether given singly or in combination (see Table 7). In contrast, the HRIG/vaccine combination and vaccine alone provided only 80% and 40% protection, respectively. We observed that, in PEP against SRV 141, the HRIG/Vaccine PEP combination appeared less successful than the vaccine alone. This result suggests that the HRIG inhibited the development of vaccine-induced immunity to the virus. In contrast, R16F7 and R14D6, by enabling 100% protection, did not apparently interfere with the potency of the vaccine. The huMabs

Table 7 Efficacy of human monoclonal antibodies against Indian dog rabies virus isolates[a]

S. No.	Group[b]	Virus	
		SRV 141	SRV 129
1	R16F7+vaccine	100%	100%
2	R14D6+vaccine	100%	100%
3	R16F7+R14D6+vaccine	100%	100%
4	HRIG+vaccine	60%	80%
5	Vaccine only	80%	40%
6	Controls	0%	0%

[a] Virus inoculation consisted of 0.05 ml administered intramuscularly in the right gastrocnemius muscle of Syrian hamsters with $\log10^{4.3}$ $MICLD_{50}/0.05$ ml Indian dog rabies virus (SRV 141) or $\log10^{4.1}$ $MICLD_{50}/0.05$ ml Indian dog rabies virus (SRV 129) (mouse-brain-passaged homogenate). Beginning 4h after virus inoculation, post-exposure prophylaxis was initiated consisting of the rabies vaccine Abhayrab at 1/10th the human dose, and administered in the left gastrocnemius muscle on days 0, 3, 7, 14, and 28. In groups 1–4 there was an additional, single administration of huMabs (either alone or in combination) and commercial human rabies immunoglobulin (Imogam, Aventis Pasteur, Lyon) (RIG at 20 IU/kg) at the site of virus inoculation on day 0. Group 5 received only vaccine. The controls, group 6, received no treatment. These experiments were done essentially as described by Hanlon et al. (2005)

appear to have a spectrum of neutralizing activity suggesting that they will be efficacious for use in at least the Indian subcontinent. Studies are underway to specifically map the binding sites of R16F7 and R14D6 on neutralization domain III and to evaluate the ability of this pair of antibodies to cross-neutralize spontaneously arising variant Indian RVs.

Our studies on molecular epidemiology indicate that the virus circulating in the Indian subcontinent belongs to GT1, and dogs are the primary transmitters of the disease. Anti-rabies antibodies against the PV strain neutralize GT1, GT4, GT5, and GT6 viruses (Badrane et al. 2001). Abhayrab incorporates a PV strain that is likely to neutralize the RVs circulating in India. Moreover, the huMabs generated using the PV strain will be appropriate for passive immunotherapy in India. Abhayrab and these monoclonal antibodies will be distributed through the Abhay clinics as well as government hospitals to ensure availability of quality products and administration. India accounts for nearly 60% of the total number of global human rabies deaths and a wider use of affordable vaccine and huMabs relevant to the Indian subcontinent will reduce the mortality rate to negligible numbers.

11   Methods for Production of Human Rabies Monoclonal Antibodies for PEP

The widespread use of ERIG and HRIG is constrained in developing countries by their limited availability and high cost. A monoclonal RIG will only be of value to the majority of people who need it if it can be produced in large amounts for a cost comparable to, and ideally less than, the cost of current ERIG and HRIG products. Initial industrial-level scale-up of heterohybridoma cell lines may not be cost-effective because of instability and low levels of antibody production. However, the cost-effective production of these huMabs and avoidance of instability of heterohybrido-mas can be achieved by taking advantage of recombinant DNA technology to clone the Ig H and -L chain genes into suitable expression vectors and expressing the inserted protein-coding sequences in appropriate cells, preferably eukaryotic cells (Morimoto et al. 2001b). The Ig heavy and light chains of several Mabs have been cloned into different expression vectors, which has enabled the expression of functional anti-bodies in a variety of cells, including lymphoid and nonlymphoid mammalian cells (Owens and Young 1994), insect cells (Liang et al. 2001), and plants (Whitelam et al. 1994; Ko et al. 2003; Girard et al. 2006). For cost-effective and industrial-scale antibody production, a rhabdovirus-based vector has shown promise and several advantages such as ease of genetic manipulation and suitability for propagation in mammalian cell culture, enabling rapid production of large amounts of antibody (Morimoto et al. 2001b; Prosniak et al. 2003). Rabies huMabs produced by this method have shown effectiveness in murine and Syrian hamster PEP models and could, if produced in accordance with industrial standards, provide an acceptable solution to the current global shortage of HRIG.

Plant expression systems may also be ideal for the economical, high-level expression of rabies huMabs. Since the initial report of functional Mabs expressed in transgenic plants (Hiatt et al. 1989), therapeutic and diagnostic Mabs have been successfully produced in transgenic tobacco (Ma et al. 1998) and other plants (Daniell et al. 2001). Plant expression systems have several advantages, such as well-established cultivation and downstream processing of plant products, easily scaled-up levels of production, and minimal risk of human pathogen and toxin contamination (Ma et al. 1998; Lerouge et al. 2000; Bakker et al. 2001; Ko et al. 2003). Antibodies made in plants are also efficiently glycosylated, although the effects of the plant-type glycosylations on the function and immunogenicity of the produced Mabs have yet to be fully characterized (Girard et al. 2006). The potential economic superiority of these systems for recombinant protein production has been extensively reviewed (Hellwig et al. 2004; Stoger et al. 2005). The effectiveness of plant expression systems for the production of functional human rabies monoclonal antibodies have been demonstrated (Dietzschold et al. 1990; Ko et al. 2003; Goudsmit et al. 2006). They used Agrobacterium-mediated transformation of the tobacco plant *Nicotiana tabacum* to express SO57, a version of the CR57 antibody used in the antibody cocktail described above. The antibody was retained in the endoplasmic reticulum and could be isolated from soluble protein extracts by standard protein A affinity purification (Ko et al. 2003). Consistent with its intracellular localization, the glycosylation of the SO57 antibody was largely composed of oligomannose-type N-glycans, but without the plant-specific α (1–3)-linked fucose residues that may engender an allergic or immune reaction in human recipients. In the Syrian hamster PEP model, the SO57 antibody given in combination with vaccine afforded 100% protection against an SRV strain, similar to the activity observed with a version of the same antibody expressed by mammalian cells (Ko et al. 2003). Cells from the transgenic tobacco plants were subsequently adapted to plant cell culture, which may complement whole-plant expression systems by enabling industrial production of antibody that does not depend on the agricultural growing season (Girard et al. 2006).

12   Conclusions and Future Challenges

In this article we have discussed the public health impact of rabies and various approaches available for effective rabies prevention and control aimed at possible reduction of rabies burden and the associated human deaths. Rabies continues to pose a significant threat to those who live in and visit endemic countries, which are mainly developing nations. A lack of sustained efforts toward meaningful reduction of vector populations, disease surveillance, and nationwide rabies control programs has resulted in persistence of the problem. Although rabies is vaccine preventable, the WHO has recommended combined administration of vaccine and RIG in cases of category III bites. The equine RIG is in wide use in developing countries despite the inherent limitations associated with heterologous RIG. However, the human RIG is preferred in developed countries for reasons of safety and efficacy. The cost, limited availability, and need for human donors have stimulated efforts to develop alternatives for HRIG that are safe, efficacious, and potentially less expensive, and we and others have created cloned human antibodies that may fill this need.

Continued ignorance on the issues related to rabies control may lead to a more complex situation than ever before. Rabies may fail to attract the attention of policy makers for allotment of sufficient funds for implementation of nationwide disease surveillance and control programs if it continues to remain a nonnotifiable disease in many developing countries. Given the changing socioeconomic conditions accompanying globalization, the inevitable human and animal translocations may lead to rabies exposures and the introduction of rabies into rabies-free countries. The continuation of human rabies exposures stresses the need to have better preparedness in terms of the availability of rabies biologicals and education on rabies. To supply quality rabies biologicals of proven potency and efficacy with a proper cold chain to the remote parts of rabies-endemic countries at affordable cost may be a real challenge. It remains to be seen whether new rabies prevention and control strategies, aided by novel rabies biologicals, will find broad use in the rabies-enzootic Third World countries and lead to a meaningful reduction in the incidence or elimination of the disease.

Acknowledgements We gratefully acknowledge M. Narendra Babu, E.V. Seshagiri, S. Rajalakshmi, R. Ramya, and P. Ramadevudu and the technical staff in the CDC Rabies program for their excellent technical support.

References

Amengual B, Whitby JE, King A, Cobo JS, Bourhy H (1997) Evolution of European bat lyssaviruses. J Gen Virol 78:2319–2328

Arya SC (1991) Acquisition of spongiform encephalopathies in India through sheep-brain rabies vaccination. Indian J Pediatr 58:563–565

Aubert MF, Masson E, Artois M, Barrat J (1994) Oral wildlife rabies vaccination field trials in Europe, with recent emphasis on France. Curr Top Microbiol Immunol 187:219–243

Badrane H, Tordo N (2001) Host switching in *Lyssavirus* history from the Chiroptera to the Carnivora orders. J Virol 75:8096–8104

Badrane H, Bahloul C, Perrin P, Tordo N (2001) Evidence of two *Lyssavirus* phylogroups with distinct pathogenicity and immunogenicity. J Virol 75:3268–3276

Baer GM, Lentz TL (1991) Rabies pathogenesis in the central nervous system. In: Baer GM (ed) The natural history of rabies. CRC Press, Boca Raton, pp 105–120

Bahloul C, Jacob Y, Tordo N, Perrin P (1998) DNA-based immunization for exploring the enlargement of immunological cross-reactivity against the lyssaviruses. Vaccine 16:417–425

Bahloul C, Ahmed SB, B'Chir BI, Kharmachi H, Hayouni el A, Dellagi K (2003) Post-exposure therapy in mice against experimental rabies: a single injection of DNA vaccine is as effective as five injections of cell culture-derived vaccine. Vaccine 22:177–184

Bahloul C, Taieb D, Diouani MF, Ahmed SB, Chtourou Y, B'Chir BI, Kharmachi H, Dellagi K (2006) Field trials of a very potent rabies DNA vaccine which induced long lasting virus neutralizing antibodies and protection in dogs in experimental conditions. Vaccine 24:1063–1072

Bakker H, Bardor M, Molthoff JW, Gomord V, Elbers I, Stevens LH, Jordi W, Lommen A, Faye L, Lerouge P, Bosch D (2001) Galactose-extended glycans of antibodies produced by transgenic plants. Proc Natl Acad Sci U S A 98:2899–2904

Balthazard M, Bhamanyar M, Ghodssi M, Sabeti A, Gajdusek C, Rouzbehi E (1955) Essai pratique du serum antirabique chez les mordus par loups enrages. Bull World Health Organ 13:747–772

Barth R, Gruschkau H, Bijok U, Hilfenhaus J, Hinz J, Milcke L, Moser H, Jaeger O, Ronneberger H, Weinmann E (1984) A new inactivated tissue culture rabies vaccine for use in man. Evaluation of PCEC-vaccine by laboratory tests. J Biol Stand 12:29–46

Belotto A, Leanes LF, Schneider MC, Tamayo H, Correa E (2005) Overview of rabies in the Americas. Virus Res 111:5–12

Benmansour A, Leblois H, Coulon P, Tuffereau C, Gaudin Y, Flamand A, Lafay F (1991) Antigenicity of rabies virus glycoprotein. J Virol 65:4198–4203

Bhatia R, Ichhpujani RL, Madhusudana SN, Hemachudha T (2004) Rabies in South and Southeast Asia. In: Program and abstracts of the WHO Expert Consultation on Rabies, Geneva

Biswas S, Reddy GS, Srinivasan VA, Rangarajan PN (2001) Preexposure efficacy of a novel combination DNA and inactivated rabies virus vaccine. Hum Gene Ther 12:1917–1922

Blancou J (1988) Ecology and epidemiology of fox rabies. Rev Infect Dis 10 [Suppl 4]: S606–S609

Bogel K, Motschwiller E (1986) Incidence of rabies and post-exposure treatment in developing countries. Bull World Health Organ 64:883–887

Bourhy H, Kissi B, Lafon M, Sacramento D, Tordo N (1992) Antigenic and molecular characterization of bat rabies virus in Europe. J Clin Microbiol 30:2419–2426

Bourhy H, Kissi B, Tordo N (1993) Molecular diversity of the *Lyssavirus* genus. Virology 194:70–81

Bregenholt S, Haurum J (2004) Pathogen-specific recombinant human polyclonal antibodies: biodefence applications. Expert Opin Biol Ther 4:387–396

Brekke OH, Sandlie I (2003) Therapeutic antibodies for human diseases at the dawn of the twenty-first century. Nat Rev Drug Discov 2:52–62

Briggs DJ, Banzhoff A, Nicolay U, Sirikwin S, Dumavibhat B, Tongswas S, Wasi C (2000) Antibody response of patients after postexposure rabies vaccination with small intradermal doses of purified chick embryo cell vaccine or purified Vero cell rabies vaccine. Bull World Health Organ 78:693–698

Brookes SM, Parsons G, Johnson N, McElhinney LM, Fooks AR (2005) Rabies human diploid cell vaccine elicits cross-neutralising and cross-protecting immune responses against European and Australian bat lyssaviruses. Vaccine 23:4101–4109

Carroll WL, Thielemans K, Dilley J, Levy R (1986) Mouse x human heterohybridomas as fusion partners with human B cell tumors. J Immunol Methods 89:61–72

Casadevall A (2002) Passive antibody administration (immediate immunity) as a specific defense against biological weapons. Emerg Infect Dis 8:833–841

Center for Biologics Evaluation and Research (1995) Guidance for industry: considerations for plasmid DNA vaccines for infectious disease indications. http://www.fda.gov/cber/gdlns/plasdnavac.pdf. Cited 18 May 2007

Centers for Disease Control and Prevention (1999) Human rabies prevention—United States, 1999. Recommendations of the Advisory Committee on Immunization Practices. MMWR Recomm Rep 48:1–21

Centers for Disease Control (2004) Manufacturer's recall of human rabies vaccine. MMWR Dispatch 53:287–289

Champion JM, Kean RB, Rupprecht CE, Notkins AL, Koprowski H, Dietzschold B, Hooper DC (2000) The development of monoclonal human rabies virus-neutralizing antibodies as a substitute for pooled human immune globulin in the prophylactic treatment of rabies virus exposure. J Immunol Methods 235:81–90

Cheung SC, Dietzschold B, Koprowski H, Notkins AL, Rando RF (1992) A recombinant human Fab expressed in Escherichia coli neutralizes rabies virus. J Virol 66:6714–6720

Childs JE, Curns AT, Dey ME, Real LA, Feinstein L, Bjornstad ON, Krebs JW (2000) Predicting the local dynamics of epizootic rabies among raccoons in the United States. Proc Natl Acad Sci U S A 97:13666–13671

Chin J (2000) Control of communicable diseases manual. APHA, Washington, DC

Chutivongse S, Wilde H, Supich C, Baer GM, Fishbein DB (1990) Postexposure prophylaxis for rabies with antiserum and intradermal vaccination. Lancet 335:896–898

Chutivongse S, Wilde H, Fishbein DB, Baer GM, Hemachudha T (1991) One-year study of the 2-1-1 intramuscular postexposure rabies vaccine regimen in 100 severely exposed Thai patients using rabies immune globulin and Vero cell rabies vaccine. Vaccine 9:573–576

Cleaveland S, Kaare M, Tiringa P, Mlengeya T, Barrat J (2003) A dog rabies vaccination campaign in rural Africa: impact on the incidence of dog rabies and human dog-bite injuries. Vaccine 21:1965–1973

Cliquet F, Aubert M, Sagne L (1998) Development of a fluorescent antibody virus neutralisation test (FAVN test) for the quantitation of rabies-neutralising antibody. J Immunol Methods 212:79–87

Coleman PG, Fevre EM, Cleaveland S (2004) Estimating the public health impact of rabies. Emerg Infect Dis 10:140–142

Corn JL, Mendez JR, Catalan EE (2003) Evaluation of baits for delivery of oral rabies vaccine to dogs in Guatemala. Am J Trop Med Hyg 69:155–158

Cox JH, Dietzschold B, Schneider LG (1977) Rabies virus glycoprotein. II. Biological and sero-logical characterization. Infect Immun 16:754–759

Daniell H, Streatfield SJ, Wycoff K (2001) Medical molecular farming: production of antibodies, biopharmaceuticals and edible vaccines in plants. Trends Plant Sci 6:219–226

David D, Yakobson BA, Gershkovich L, Gayer S (2004) Tracing the regional source of rabies infection in an Israeli dog by viral analysis. Vet Rec 155:496–497

de Kruif J, Bakker AB, Marissen WE, Kramer RA, Throsby M, Rupprecht CE, Goudsmit J (2007) A human monoclonal antibody cocktail as a novel component of rabies postexposure prophylaxis. Annu Rev Med 58:359–368

Dean DJ, Baer GM, Thompson WR (1963) Studies on the local treatment of rabies-infected wounds. Bull World Health Organ 28:477–486

Dietzschold B, Schnell MJ (2002) New approaches to the development of live attenuated rabies vaccines. Hybrid Hybridomics 21:129–134

Dietzschold B, Tollis M, Lafon M, Wunner WH, Koprowski H (1987) Mechanisms of rabies virus neutralization by glycoprotein-specific monoclonal antibodies. Virology 161:29–36

Dietzschold B, Gore M, Casali P, Ueki Y, Rupprecht CE, Notkins AL, Koprowski H (1990) Biological characterization of human monoclonal antibodies to rabies virus. J Virol 64: 3087–3090

Dietzschold B, Kao M, Zheng YM, Chen ZY, Maul G, Fu ZF, Rupprecht CE, Koprowski H (1992) Delineation of putative mechanisms involved in antibody-mediated clearance of rabies virus from the central nervous system. Proc Natl Acad Sci U S A 89:7252–7256

Dietzschold B, Faber M, Schnell MJ (2003) New approaches to the prevention and eradication of rabies. Expert Rev Vaccines 2:399–406

Dreesen DW, Fishbein DB, Kemp DT, Brown J (1989) Two-year comparative trial on the immunogenicity and adverse effects of purified chick embryo cell rabies vaccine for pre-exposure immunization. Vaccine 7:397–400

Enssle K, Kurrle R, Kohler R, Muller H, Kanzy EJ, Hilfenhaus J, Seiler FR (1991) A rabies-specific human monoclonal antibody that protects mice against lethal rabies. Hybridoma 10:547–556

Ertl HC (2005) Immunological insights from genetic vaccines. Virus Res 111:89–92

Ertl HC, Xiang ZQ (1996) Genetic immunization. Viral Immunol 9:1–9

Estrada R, Vos A, De Leon R, Mueller T (2001) Field trial with oral vaccination of dogs against rabies in the Philippines. BMC Infect Dis 1:23

Farrugia A, Poulis P (2001) Intravenous immunoglobulin: regulatory perspectives on use and supply. Transfus Med 11:63–74

Fu ZF (1997) Rabies and rabies research: past, present and future. Vaccine 15 Suppl:S20–S24

Fuenzalida E, Palacios R (1955) An improved method for preparation of rabies vaccine [in Spanish]. Biol Inst Bacteriol Chile 8:3–10

Funaro A, Horenstein AL, Ghisolfi G, Bussolati B, Bartorelli A, Bussolati G (1999) Identification of a 220-kDa membrane tumor-associated antigen by human anti-UK114 monoclonal antibodies selected from the immunoglobulin repertoire of a cancer patient. Exp Cell Res 247:441–450

Girard LS, Fabis MJ, Bastin M, Courtois D, Petiard V, Koprowski H (2006) Expression of a human anti-rabies virus monoclonal antibody in tobacco cell culture. Biochem Biophys Res Commun 345:602–607

Goudsmit J, Marissen WE, Weldon WC, Niezgoda M, Hanlon CA, Rice AB, Kruif J, Dietzschold B, Bakker AB, Rupprecht CE (2006) Comparison of an anti-rabies human monoclonal antibody combination with human polyclonal anti-rabies immune globulin. J Infect Dis 193:796–801

Haddad N, Ben Khelifa R, Matter H, Kharmachi H, Aubert MF, Wandeler A, Blancou J (1994) Assay of oral vaccination of dogs against rabies in Tunisia with the vaccinal strain SADBern. Vaccine 12:307–309

Hanlon CA, Niezgoda M, Morrill PA, Rupprecht CE (2001) The incurable wound revisited: progress in human rabies prevention? Vaccine 19:2273–2279

Hassett DE, Whitton JL (1996) DNA immunization. Trends Microbiol 4:307–312

Hattwick MA, Corey L, Creech WB (1976) Clinical use of human globulin immune to rabies virus. J Infect Dis 133 Suppl:A266–272

Haurum J, Bregenholt S (2005) Recombinant polyclonal antibodies: therapeutic antibody technologies come full circle. IDrugs 8:404–409

Hellwig S, Drossard J, Twyman RM, Fischer R (2004) Plant cell cultures for the production of recombinant proteins. Nat Biotechnol 22:1415–1422

Hemachuda T (1989) Rabies. In: Vinken PJ, Bruyn GW, Klawans HL, McKendall RR (eds) Viral disease. Elsevier, Amsterdam, pp 383–404

Hemachuda T (1994) Human rabies: clinical aspects, pathogenesis, and potential therapy. Curr Top Microbiol Immunol 187:121–143

Hemachuda T, Mitrabhakdi E (2000) Rabies. In: Davis LE, Kennedy PGE (eds) Infectious diseases of the nervous system. Butterworth Heinemann, Oxford, pp 401–444

Hemachuda T, Phuapradit P (1997) Rabies. Curr Opin Neurol 10:260–267

Hemachuda T, Phanuphak P, Johnson RT, Griffin DE, Ratanavongsiri J, Siriprasomsup W (1987) Neurologic complications of Semple-type rabies vaccine: clinical and immunologic studies. Neurology 37:550–556

Hemachuda T, Laothamatas J, Rupprecht CE (2002) Human rabies: a disease of complex neuropathogenetic mechanisms and diagnostic challenges. Lancet Neurol 1:101–109

Hiatt A, Cafferkey R, Bowdish K (1989) Production of antibodies in transgenic plants. Nature 342:76–78

Holmes EC, Woelk CH, Kassis R, Bourhy H (2002) Genetic constraints and the adaptive evolution of rabies virus in nature. Virology 292:247–257

Hooper DC, Morimoto K, Bette M, Weihe E, Koprowski H, Dietzschold B (1998) Collaboration of antibody and inflammation in clearance of rabies virus from the central nervous system. J Virol 72:3711–3719

Hooper PT, Lunt RA, Gould AR, Samaratunga H, Hyatt AD, Gleeson LJ, Rodwell BJ, Rupprecht CE, Smith JS, Murray PK (1997) A new lyssavirus—the first endemic rabies-related virus recognized in Australia. Bull Inst Pasteur 95:209–218

Ichhpujani RL, Bhardwaj M, Chhabra M, Datta KK (2001) Rabies in humans in India. In: Dodet B, Meslin FX, Haseltine E (eds) Proceedings of the fourth international symposium on rabies control in Asia. John Libbey Eurotext, London, pp 212–213

Ito M, Itou T, Shoji Y, Sakai T, Ito FH, Arai YT, Takasaki T, Kurane I (2003) Discrimination between dog-related and vampire bat-related rabies viruses in Brazil by strain-specific reverse transcriptase-polymerase chain reaction and restriction fragment length polymorphism analysis. J Clin Virol 26:317–330

Jallet C, Jacob Y, Bahloul C, Drings A, Desmezieres E, Tordo N, Perrin P (1999) Chimeric lyssavirus glycoproteins with increased immunological potential. J Virol 73:225–233

Jenkins SR, Perry BD, Winkler WG (1988) Ecology and epidemiology of raccoon rabies. Rev Infect Dis 10 [Suppl 4]:S620–S625

Jenkins SR, Auslander M, Conti L, Leslie MJ, Sorhage FE, Sun B (2004) Compendium of animal rabies prevention and control, 2004. J Am Vet Med Assoc 224:216–222

Kallel H, Diouani MF, Loukil H, Trabelsi K, Snoussi MA, Majoul S, Rourou S, Dellagi K (2006) Immunogenicity and efficacy of an in-house developed cell-culture derived veterinarian rabies vaccine. Vaccine 24:4856–4862

Khawplod P, Wilde H, Chomchey P, Benjavongkulchai M, Yenmuang W, Chaiyabutr N, Sitprija V (1996) What is an acceptable delay in rabies immune globulin administration when vaccine alone had been given previously? Vaccine 14:389–391

Kilic B, Unal B, Semin S, Konakci SK (2006) An important public health problem: rabies suspected bites and post-exposure prophylaxis in a health district in Turkey. Int J Infect Dis 10:248–254

Kitala PM, McDermott JJ, Kyule MN, Gathuma JM (2000) Community-based active surveillance for rabies in Machakos District, Kenya. Prev Vet Med 44:73–85

Knobel DL, Cleaveland S, Coleman PG, Fevre EM, Meltzer MI, Miranda ME, Shaw A, Zinsstag J, Meslin FX (2005) Re-evaluating the burden of rabies in Africa and Asia. Bull World Health Organ 83:360–368

Ko K, Tekoah Y, Rudd PM, Harvey DJ, Dwek RA, Spitsin S, Hanlon CA, Rupprecht C, Dietzschold B, Golovkin M, Koprowski H (2003) Function and glycosylation of plant-derived antiviral monoclonal antibody. Proc Natl Acad Sci U S A 100:8013–8018

Koprowski H, Van Der Scheer J, Black J (1950) Use of hyperimmune anti-rabies serum concentrates in experimental rabies. Am J Med 8:412–420

Koprowski H, Wiktor T, Abelseth M (1985) Cross-reactivity and cross-protection: rabies variants and rabies-related viruses. In: Kuwert E, Merieux C, Koprowski H, Bogel K (eds) Rabies in the tropics. Springer-Verlag, Berlin Heidelberg New York, pp 30–39

Kramer RA, Marissen WE, Goudsmit J, Visser TJ, Clijsters-Van der Horst M, Bakker AQ, de Jong M, Jongeneelen M, Thijsse S, Backus HH, Rice AB, Weldon WC, Rupprecht CE, Dietzschold B, Bakker AB, de Kruif J (2005) The human antibody repertoire specific for rabies virus glycoprotein as selected from immune libraries. Eur J Immunol 35:2131–2145

Krebs JW, Mandel EJ, Swerdlow DL, Rupprecht CE (2005) Rabies surveillance in the United States during 2004. J Am Vet Med Assoc 227:1912–1925

Kureishi A, Xu LZ, Wu H, Stiver HG (1992) Rabies in China: recommendations for control. Bull World Health Organ 70:443–450

Kuzmin IV, Hughes GJ, Botvinkin AD, Orciari LA, Rupprecht CE (2005) Phylogenetic relationships of Irkut and West Caucasian bat viruses within the Lyssavirus genus and suggested quantitative criteria based on the N gene sequence for Lyssavirus genotype definition. Virus Res 111:28–43

Lafon M, Edelman L, Bouvet JP, Lafage M, Montchatre E (1990) Human monoclonal antibodies specific for the rabies virus glycoprotein and N protein. J Gen Virol 71:1689–1696

Lerouge P, Bardor M, Pagny S, Gomord V, Faye L (2000) N-Glycosylation of recombinant pharmaceutical glycoproteins produced in transgenic plants: towards an humanisation of plant N-glycans. Curr Pharm Biotechnol 1:347–354

Liang M, Dubel S, Li D, Queitsch I, Li W, Bautz EK (2001) Baculovirus expression cassette vectors for rapid production of complete human IgG from phage display selected antibody fragments. J Immunol Methods 247:119–130

Libeau G, Lafon M (1984) Production of monoclonal antibodies against the Pasteur (P.V.) strain of rabies virus: problems and results. Dev Biol Stand 57:213–218

Lodmell DL, Ewalt LC (2001) Post-exposure DNA vaccination protects mice against rabies virus. Vaccine 19:2468–2473

Lodmell DL, Ray NB, Parnell MJ, Ewalt LC, Hanlon CA, Shaddock JH, Sanderlin DS, Rupprecht CE (1998) DNA immunization protects nonhuman primates against rabies virus. Nat Med 4:949–952

Lodmell DL, Ray NB, Ulrich JT, Ewalt LC (2000) DNA vaccination of mice against rabies virus: effects of the route of vaccination and the adjuvant monophosphoryl lipid A (MPL). Vaccine 18:1059–1066

Lodmell DL, Parnell MJ, Bailey JR, Ewalt LC, Hanlon CA (2001) One-time gene gun or intramuscular rabies DNA vaccination of non-human primates: comparison of neutralizing antibody responses and protection against rabies virus 1 year after vaccination. Vaccine 20:838–844

Ma JK, Hikmat BY, Wycoff K, Vine ND, Chargelegue D, Yu L, Hein MB, Lehner T (1998) Characterization of a recombinant plant monoclonal secretory antibody and preventive immunotherapy in humans. Nat Med 4:601–606

MacInnes CD, Smith SM, Tinline RR, Ayers NR, Bachmann P, Ball DG, Calder LA, Crosgrey SJ, Fielding C, Hauschildt P, Honig JM, Johnston DH, Lawson KF, Nunan CP, Pedde MA, Pond B, Stewart RB, Voigt DR (2001) Elimination of rabies from red foxes in eastern Ontario. J Wildl Dis 37:119–132

Marissen WE, Kramer RA, Rice A, Weldon WC, Niezgoda M, Faber M, Slootstra JW, Meloen RH, Clijsters-van der Horst M, Visser TJ, Jongeneelen M, Thijsse S, Throsby M, de Kruif J, Rupprecht CE, Dietzschold B, Goudsmit J, Bakker AB (2005) Novel rabies virus-neutralizing epitope recognized by human monoclonal antibody: fine mapping and escape mutant analysis. J Virol 79:4672–4678

Martinez L (2000) Global infectious disease surveillance. Int J Infect Dis 4:222–228

Matha IS, Salunke SR (2005) Immunogenicity of purified vero cell rabies vaccine used in the treatment of fox-bite victims in India. Clin Infect Dis 40:611–613

May RM, Anderson RM (1984) Spatial heterogeneity and the design of immunization programs. Math Biosci 72:83–111

Mayen F (2003) Haematophagous bats in Brazil, their role in rabies transmission, impact on public health, livestock industry and alternatives to an indiscriminate reduction of bat population. J Vet Med B Infect Dis Vet Public Health 50:469–472

McKay N, Wallis L (2005) Rabies: a review of UK management. Emerg Med J 22:316–321

Mebatsion T, Weiland F, Conzelmann KK (1999) Matrix protein of rabies virus is responsible for the assembly and budding of bullet-shaped particles and interacts with the transmembrane spike glycoprotein G. J Virol 73:242–250

Meslin FX, Kaplan MM (1996) General considerations in the production and use of brain-tissue and purified chicken embryo rabies vaccines for human use. In: Meslin FX, Kaplan MM, koprowski H (eds) Laboratory techniques in rabies. WHO, Geneva, pp 221–233

Meslin FX, Fishbein DB, Matter HC (1994) Rationale and prospects for rabies elimination in developing countries. Curr Top Microbiol Immunol 187:1–26

Messenger SL, Smith JS, Rupprecht CE (2002) Emerging epidemiology of bat-associated cryptic cases of rabies in humans in the United States. Clin Infect Dis 35:738–747

Mitmoonpitak C, Tepsumethanon V, Wilde H (1998) Rabies in Thailand. Epidemiol Infect 120:165–169

Montagnon B, Vincent-Falquet JC, Fanget B (1983) Thousand litre scale microcarrier culture of Vero cells for killed polio virus vaccine. Promising results. Dev Biol Stand 55:37–42

Morimoto K, Patel M, Corisdeo S, Hooper DC, Fu ZF, Rupprecht CE, Koprowski H, Dietzschold B (1996) Characterization of a unique variant of bat rabies virus responsible for newly emerging human cases in North America. Proc Natl Acad Sci U S A 93:5653–5658

Morimoto K, McGettigan JP, Foley HD, Hooper DC, Dietzschold B, Schnell MJ (2001a) Genetic engineering of live rabies vaccines. Vaccine 19:3543–3551

Morimoto K, Schnell MJ, Pulmanausahakul R, McGettigan JP, Foley HD, Faber M, Hooper DC, Dietzschold B (2001b) High level expression of a human rabies virus-neutralizing monoclonal antibody by a rhabdovirus-based vector. J Immunol Methods 252:199–206

Mudur G (2005) Foreign visitors to India are unaware of rabies risk. BMJ 331:255

Muller BH, Lafay F, Demangel C, Perrin P, Tordo N, Flamand A, Lafaye P, Guesdon JL (1997) Phage-displayed and soluble mouse scFv fragments neutralize rabies virus. J Virol Methods 67:221–233

Nadin-Davis SA, Abdel-Malik M, Armstrong J, Wandeler AI (2002) *Lyssavirus* P gene characterisation provides insights into the phylogeny of the genus and identifies structural similarities and diversity within the encoded phosphoprotein. Virology 298:286–305

Nagarajan T, Mohanasubramanian B, Seshagiri EV, Nagendrakumar SB, Saseendranath MR, Satyanarayana ML, Thiagarajan D, Rangarajan PN, Srinivasan VA (2006a) Molecular epidemiology of rabies virus isolates in India. J Clin Microbiol 44:3218–3224

Nagarajan T, Reddy GS, Mohana Subramanian B, Rajalakshmi S, Thiagarajan D, Tordo N, Jallet C, Srinivasan VA (2006b) A simple immuno-capture ELISA to estimate rabies viral glycoprotein antigen in vaccine manufacture. Biologicals 34:21–27

Nel LH, Sabeta CT, von Teichman B, Jaftha JB, Rupprecht CE, Bingham J (2005) Mongoose rabies in southern Africa: a re-evaluation based on molecular epidemiology. Virus Res 109:165–173

Nicholson KG (1996) Cell-culture vaccines for human use: general considerations. In: Meslin FX, Kaplan MM, Koprowski H (eds) Laboratory techniques in rabies. WHO, Geneva, pp 271–279

Nishizono A, Mannen K, Elio-Villa LP, Tanaka S, Li KS, Mifune K, Arca BF, Cabanban A, Martinez B, Rodriguez A, Atienza VC, Camba R, Resontoc N (2002) Genetic analysis of rabies virus isolates in the Philippines. Microbiol Immunol 46:413–417

Owens RJ, Young RJ (1994) The genetic engineering of monoclonal antibodies. J Immunol Methods 168:149–165

Parviz S, Luby S, Wilde H (1998) Postexposure treatment of rabies in Pakistan. Clin Infect Dis 27:751–756

Pasquini S, Deng H, Reddy ST, Giles-Davis W, Ertl HC (1999) The effect of CpG sequences on the B cell response to a viral glycoprotein encoded by a plasmid vector. Gene Ther 6:1448–1455

Patrick GR, O'Rourke KM (1998) Dog and cat bites: epidemiologic analyses suggest different prevention strategies. Public Health Rep 113:252–257

Perera MA, Harischandra PA, Wimalaratne O, Damboragama SN (2000) Feasibility of canine oral rabies vaccination in Sri Lanka—a preliminary report. Ceylon Med J 45:61–64

Perrin P, Madhusudana S, Gontier-Jallet C, Petres S, Tordo N, Merten OW (1995) An experimental rabies vaccine produced with a new BHK-21 suspension cell culture process: use of serum-free medium and perfusion-reactor system. Vaccine 13:1244–1250

Perrin P, Jacob Y, Aguilar-Setien A, Loza-Rubio E, Jallet C, Desmezieres E, Aubert M, Cliquet F, Tordo N (1999) Immunization of dogs with a DNA vaccine induces protection against rabies virus. Vaccine 18:479–486

Pinto AR, Reyes-Sandoval A, Ertl HC (2003) Chemokines and TRANCE as genetic adjuvants for a DNA vaccine to rabies virus. Cell Immunol 224:106–113

Plotkin SA (1980) Rabies vaccine prepared in human cell cultures: progress and perspectives. Rev Infect Dis 2:433–448

Plotkin SA, Rupprecht CE, Koprowski H (1999) Rabies vaccine. In: Plotkin SA, Orenstein WA (eds) Vaccines. WB Saunders, Philadelphia, pp 743–766

Prehaud C, Coulon P, LaFay F, Thiers C, Flamand A (1988) Antigenic site II of the rabies virus glycoprotein: structure and role in viral virulence. J Virol 62:1–7

Prosniak M, Faber M, Hanlon CA, Rupprecht CE, Hooper DC, Dietzschold B (2003) Development of a cocktail of recombinant-expressed human rabies virus-neutralizing monoclonal antibodies for postexposure prophylaxis of rabies. J Infect Dis 188:53–56

Ramanna BC, Reddy GS, Srinivasan VA (1991) An outbreak of rabies in cattle and use of tissue culture rabies vaccine during the outbreak. J Commun Dis 23:283–285

Rath A, Choudhury S, Batra D, Kapre SV, Rupprecht CE, Gupta SK (2005) DNA vaccine for rabies: relevance of the transmembrane domain of the glycoprotein in generating an antibody response. Virus Res 113:143–152

Ray NB, Ewalt LC, Lodmell DL (1997) Nanogram quantities of plasmid DNA encoding the rabies virus glycoprotein protect mice against lethal rabies virus infection. Vaccine 15:892–895

Real LA, Henderson JC, Biek R, Snaman J, Jack TL, Childs JE, Stahl E, Waller L, Tinline R, Nadin-Davis S (2005) Unifying the spatial population dynamics and molecular evolution of epidemic rabies virus. Proc Natl Acad Sci U S A 102:12107–12111

Rotz LD, Hensley JA, Rupprecht CE, Childs JE (1998) Large-scale human exposures to rabid or presumed rabid animals in the United States: 22 cases (1990–1996). J Am Vet Med Assoc 212:1198–1200

Rupprecht CE, Gibbons RV (2004) Clinical practice. Prophylaxis against rabies. N Engl J Med 351:2626–2635

Rupprecht CE, Charlton KM, Artois M, Casey GA, Webster WA, Campbell JB, Lawson KF, Schneider LG (1990) Ineffectiveness and comparative pathogenicity of attenuated rabies virus vaccines for the striped skunk (Mephitis mephitis). J Wildl Dis 26:99–102

Rupprecht CE, Hanlon CA, Cummins LB, Koprowski H (1992) Primate responses to a vaccinia-rabies glycoprotein recombinant virus vaccine. Vaccine 10:368–374

Rupprecht CE, Hanlon CA, Hemachudha T (2002) Rabies re-examined. Lancet Infect Dis 2:327–343

Rupprecht CE, Hanlon CA, Blanton J, Manangan J, Morrill P, Murphy S, Niezgoda M, Orciari LA, Schumacher CL, Dietzschold B (2005) Oral vaccination of dogs with recombinant rabies virus vaccines. Virus Res 111:101–105

Sampath G, Reddy SV, Rao ML, Rao YU, Palaniappan C (2005) An immunogenicity study of a newly introduced purified Vero cell rabies vaccine (Abhayrab) manufactured in India. Vaccine 23:897–900

Schumacher CL, Dietzschold B, Ertl HC, Niu HS, Rupprecht CE, Koprowski H (1989) Use of mouse anti-rabies monoclonal antibodies in postexposure treatment of rabies. J Clin Invest 84:971–975

Servat A, Lutsch C, Delore V, Lang J, Veitch K, Cliquet F (2003) Efficacy of rabies immunoglobulins in an experimental post-exposure prophylaxis rodent model. Vaccine 22: 244–249

Shah U, Jaswal GS (1975) Follow-up of rabid dog bite victims. Lancet 2:653–655

Shetty RA, Chaturvedi S, Singh Z (2005) Profile of animal bite cases in Pune. J Commun Dis 37:66–72

Siemoneit K, da Silva Cardoso M, Wolpl A, Koerner K, Subanek B (1994) Isolation and epitope characterization of human monoclonal antibodies to hepatitis C virus core antigen. Hybridoma 13:9–13

Sikes RK, Peacock GV, Acha P, Arko RJ, Dierks R (1971) Rabies vaccines: duration-of-immunity study in dogs. J Am Vet Med Assoc 159:1491–1499

Sitthi-Amorn C, Jiratanavattana V, Keoyoo J, Sonpunya N (1987) The diagnostic properties of laboratory tests for rabies. Int J Epidemiol 16:602–605

Slade HB (1994) Human immunoglobulins for intravenous use and hepatitis C viral transmission. Clin Diagn Lab Immunol 1:613–619

Smith JS, Fishbein DB, Rupprecht CE, Clark K (1991) Unexplained rabies in three immigrants in the United States. A virologic investigation. N Engl J Med 324:205–211

Steck F, Wandeler A, Bichsel P, Capt S, Hafliger U, Schneider L (1982) Oral immunization of foxes against rabies. Laboratory and field studies. Comp Immunol Microbiol Infect Dis 5:165–171

Steele JH (1988) Rabies in the Americas and remarks on global aspects. Rev Infect Dis 10 [Suppl 4]: S585–597

Stoger E, Sack M, Nicholson L, Fischer R, Christou P (2005) Recent progress in plantibody technology. Curr Pharm Des 11:2439–2457

Stohr K, Meslin FM (1996) Progress and setbacks in the oral immunisation of foxes against rabies in Europe. Vet Rec 139:32–35

Strady A, Lang J, Rotivel Y, Jaussaud R, Fritzell C, Tsiang H (1996) [Immunoprophylaxis of rabies: current recommendations]. Presse Med 25:1023–1027

Sudarshan MK, Mahendra BJ, Narayan DH (2001) A community survey of dog bites, anti-rabies treatment, rabies and dog population management in Bangalore city. J Commun Dis 33:245–251

Sudarshan MK, Madhusudana SN, Mahendra BJ, Rao NS, Ashwath Narayana DH, Abdul Rahman S, Meslin FX, Lobo D, Ravikumar K, Gangaboraiah (2007) Assessing the burden of human rabies in India: results of a national multi-center epidemiological survey. Int J Infect Dis 11:29–35

Suntharasamai P, Chaiprasithikul P, Wasi C, Supanaranond W, Auewarakul P, Chanthavanich P, Supapochana A, Areeraksa S, Chittamas S, Jittapalapongsa S, et al (1994) A simplified and economical intradermal regimen of purified chick embryo cell rabies vaccine for postexposure prophylaxis. Vaccine 12:508–512

Swaddiwuthipong W, Weniger BG, Wattanasri S, Warrell MJ (1988) A high rate of neurological complications following Semple anti-rabies vaccine. Trans R Soc Trop Med Hyg 82: 472–475

Taylor LH, Latham SM, Woolhouse ME (2001) Risk factors for human disease emergence. Philos Trans R Soc Lond B Biol Sci 356:983–989

Tollis M, Dietzschold B, Volia CB, Koprowski H (1991) Immunization of monkeys with rabies ribonucleoprotein (RNP) confers protective immunity against rabies. Vaccine 9:134–136

Tordo N, Poch O, Ermine A, Keith G, Rougeon F (1986) Walking along the rabies genome: is the large G-L intergenic region a remnant gene? Proc Natl Acad Sci U S A 83:3914–3918

Trabelsi K, Rourou S, Loukil H, Majoul S, Kallel H (2006) Optimization of virus yield as a strategy to improve rabies vaccine production by Vero cells in a bioreactor. J Biotechnol 121:261–271

Ueki Y, Goldfarb IS, Harindranath N, Gore M, Koprowski H, Notkins AL, Casali P (1990) Clonal analysis of a human antibody response. Quantitation of precursors of antibody-producing cells and generation and characterization of monoclonal IgM, IgG, and IgA to rabies virus. J Exp Med 171:19–34

Ulmer JB, Liu MA (2002) Ethical issues for vaccines and immunization. Nat Rev Immunol 2:291–296

Velasco-Villa A, Orciari LA, Souza V, Juarez-Islas V, Gomez-Sierra M, Castillo A, Flisser A, Rupprecht CE (2005) Molecular epizootiology of rabies associated with terrestrial carnivores in Mexico. Virus Res 111:13–27

Velasco-Villa A, Orciari LA, Juarez-Islas V, Gomez-Sierra M, Padilla-Medina I, Flisser A, Souza V, Castillo A, Franka R, Escalante-Mane M, Sauri-Gonzalez I, Rupprecht CE (2006) Molecular diversity of rabies viruses associated with bats in Mexico and other countries of the Americas. J Clin Microbiol 44:1697–1710

Wandeler A (2004) Epidemiology and ecology of fox rabies in Europe. In: King AA, Fooks AR, Aubert M, Wandeler A (eds) Historical perspective of rabies in Europe and the Mediterranean basin. OIE, Paris, pp 201–214

Wandeler AI, Budde A, Capt S, Kappeler A, Matter H (1988) Dog ecology and dog rabies control. Rev Infect Dis 10 [Suppl 4]:S684–S688

Wandeler AI, Matter HC, Kappeler A, Budde A (1993) The ecology of dogs and canine rabies: a selective review. Rev Sci Tech 12:51–71

Wang Y, Xiang Z, Pasquini S, Ertl HC (1997) Immune response to neonatal genetic immunization. Virology 228:278–284

Wang Y, Xiang Z, Pasquini S, Ertl HC (1998) Effect of passive immunization or maternally transferred immunity on the antibody response to a genetic vaccine to rabies virus. J Virol 72:1790–1796

Warrell MJ, Nicholson KG, Warrell DA, Suntharasamai P, Chanthavanich P, Viravan C, Sinhaseni A, Chiewbambroongkiat MK, Pouradier-Duteil X, Xueref C, et al (1985) Economical multiple-site intradermal immunisation with human diploid-cell-strain vaccine is effective for postexposure rabies prophylaxis. Lancet 1:1059–1062

Whitelam GC, Cockburn W, Owen MR (1994) Antibody production in transgenic plants. Biochem Soc Trans 22:940–944

WHO (1993) Report of the fourth WHO consultation on oral immunization of dogs against rabies. WHO, Geneva. http://www.who.int/rabies/animal/en/whorabres9342.pdf. Cited 18 May 2007

WHO (1996) WHO recommendations on rabies post-exposure treatment and the correct technique of intradermal immunization. WHO, Geneva. http://whqlibdoc.who.int/hq/1996/WHO_EMC_ZOO_96.6.pdf. Cited 18 May 2007

WHO (1998) World survey of rabies N° 32 for the Year 1996. WHO, Geneva. http://whqlibdoc.who.int/hq/1998/WHO_EMC_ZDI_98.4.pdf. Cited 18 May 2007

WHO (2001) Strategies for the control and elimination of rabies in Asia. WHO, Geneva. http://www.who.int/rabies/en/Strategies_for_the_control_and_elimination_of_rabies_in_Asia.pdf. Cited 18 May 2007

WHO (2002a) Rabies vaccines. Wkly Epidemiol Rec 77:109–119

WHO (2002b) World Survey of Rabies N° 35 for the Year 1999. WHO, Geneva. http://www.who.int/rabies/resources/wsr1999/en/index.html. Cited 18 May 2007

WHO (2004) WHO expert consultation on rabies. First report, WHO technical report series 931. WHO, Geneva, p 121

WHO (2006) Rabies fact sheet N° 99. WHO, Geneva. http://www.who.int/mediacentre/factsheets/fs099/en/print.html. Cited 18 May 2007

WHO-APCRI (2004) Assessing burden of rabies in India. WHO, Geneva. http://www.apcri.org/whosurvey.pdf. Cited 18 May 2007

Wiktor TJ, Koprowski H (1978) Monoclonal antibodies against rabies virus produced by somatic cell hybridization: detection of antigenic variants. Proc Natl Acad Sci U S A 75:3938–3942

Wiktor TJ, Sokol F, Kuwert E, Koprowski H (1969) Immunogenicity of concentrated and purified rabies vaccine of tissue culture origin. Proc Soc Exp Biol Med 131:799–805

Wilde H, Choomkasien P, Hemachudha T, Supich C, Chutivongse S (1989) Failure of rabies postexposure treatment in Thailand. Vaccine 7:49–52

Wilde H, Sirikawin S, Sabcharoen A, Kingnate D, Tantawichien T, Harischandra PA, Chaiyabutr N, de Silva DG, Fernando L, Liyanage JB, Sitprija V (1996) Failure of postexposure treatment of rabies in children. Clin Infect Dis 22:228–232

Wilde H, Khawplod P, Hemachudha T, Sitprija V (2002) Postexposure treatment of rabies infection: can it be done without immunoglobulin? Clin Infect Dis 34:477–480

Wilde H, Khawplod P, Khamoltham T, Hemachudha T, Tepsumethanon V, Lumlerdacha B, Mitmoonpitak C, Sitprija V (2005) Rabies control in South and Southeast Asia. Vaccine 23:2284–2289

Willoughby RE Jr, Tieves KS, Hoffman GM, Ghanayem NS, Amlie-Lefond CM, Schwabe MJ, Chusid MJ, Rupprecht CE (2005) Survival after treatment of rabies with induction of coma. N Engl J Med 352:2508–2514

Winkler WG, Bogel K (1992) Control of rabies in wildlife. Sci Am 266:86–92

Wunner WH (2002) Rabies virus. In: Jackson AC, Wunner WH (eds) Rabies. Academic Press, San Diego, pp 23–77

Wunner WH, Larson JK, Dietzschold B, Smith CL (1988) The molecular biology of rabies viruses. Rev Infect Dis 10 [Suppl 4]:S771–S784

Xiang ZQ, Spitalnik S, Tran M, Wunner WH, Cheng J, Ertl HC (1994) Vaccination with a plasmid vector carrying the rabies virus glycoprotein gene induces protective immunity against rabies virus. Virology 199:132–140

Xiang ZQ, Spitalnik SL, Cheng J, Erikson J, Wojczyk B, Ertl HC (1995) Immune responses to nucleic acid vaccines to rabies virus. Virology 209:569–579

Yusibov V, Hooper DC, Spitsin SV, Fleysh N, Kean RB, Mikheeva T, Deka D, Karasev A, Cox S, Randall J, Koprowski H (2002) Expression in plants and immunogenicity of plant virus-based experimental rabies vaccine. Vaccine 20:3155–3164

Immunoprophylaxis of RSV Infection: Advancing from RSV-IGIV to Palivizumab and Motavizumab

H. Wu(✉), D.S. Pfarr , G.A. Losonsky, and P.A. Kiener

Abstract Antibodies mediate humoral immune responses and play key roles in the defense of viral infection by the recognition, neutralization, and elimination of viruses from the circulation. For the prevention of respiratory syncytial virus (RSV) infection, the natural immune response to RSV from pooled human plasma has been harvested and successfully developed as a prophylactic polyclonal RSV hyperimmune globulin, RespiGam (RSV-IGIV; MedImmune, Gaithersburg, MD). The success of RSV-IGIV validated the immunoprophylaxis approach for RSV prevention and led to the development of Synagis (palivizumab; MedImmune, Gaithersburg, MD), a humanized monoclonal antibody (mAb) that binds to the RSV F protein. Palivizumab is a potent anti-RSV mAb that is about 50-fold more potent than RSV-IGIV, and since obtaining regulatory approval in 1998 it has been

H. Wu

MedImmune, Inc., One MedImmune Way, Gaithersburg, MD 20878, USA

e-mail: wuh@medimmune.com

S.K. Dessain (ed.) *Human Antibody Therapeutics for Viral Disease. Current Topics in Microbiology and Immunology 317.*

used extensively to help prevent severe RSV disease in high-risk infants and children. However, a very small number of patients receiving the drug do not appear to be adequately protected. To further improve protection against RSV, we have applied a directed evolution approach to enhance the binding of palivizumab to F protein by manipulation of both the on and off rates. These efforts have yielded a more potent second-generation mAb, motavizumab, which is currently under study in phase III clinical trials. Most recently, a third generation mAb, Numax-YTE, has been generated with the intent to extend the serum half-life of the mAb in humans. If successfully developed, this drug may offer the opportunity for less frequent dosing, obviating the need for the monthly treatments that are required with palivizumab. The development of these anti-RSV approaches exemplifies the accelerated pace of drug development made possible with cutting-edge antibody engineering technologies.

1 Introduction

Respiratory syncytial virus (RSV) is a nonsegmented, negative-strand RNA virus and a member of the *Paramyxoviridae* family. First identified in 1956 (Blount et al. 1956), it was very quickly recognized as a major cause of serious respiratory tract disease in the pediatric population (reviewed by Collins et al. 2001). Premature infants and children with chronic lung disease or congenital heart disease are at high risk of severe RSV infection. Primary infection typically occurs in over 65% of children before 12 months of age (Glezen et al. 1986). However, exposure does not confer long-term immunity, and individuals are subject to reinfection during subsequent seasonal outbreaks. In addition, RSV infects the elderly and immuno-compromised individuals and can cause severe lower respiratory tract disease (Garvie and Gray 1980; Falsey and Walsh 2000).

The morbidity and mortality associated with RSV disease in very young infants, coupled with its frequency and worldwide distribution, have made it a prime target for the development of a vaccine that could be used early in life. However, to date there is no licensed RSV vaccine. A formalin-inactivated RSV vaccine used in the 1960s not only failed to induce a protective neutralizing antibody response, but also led to an increased incidence of lower respiratory tract RSV disease and death in some immunized children upon subsequent natural infection with RSV (Kapikian et al. 1969). Alternative approaches that use attenuated live virus vaccines have yet to yield, in this young patient population, the appropriate balance of sufficient infectivity to provoke a strong immune response but appropriate attenuation to minimize reactogenicity (Wright et al. 1976, 1982). Another factor leading to the difficulty of developing an RSV vaccine may be that neonates may not mount a robust protective immune response following vaccination, due to immunologic immaturity, or that the presence of circulating maternally derived anti-RSV antibodies leads to the suppression of an effective immune response (Crowe 2001).

Although the development of an effective vaccine has remained elusive, several studies have suggested that high titers of specific neutralizing antibody to RSV may provide protection to both primary and recurrent infection. It has been shown that the severity of RSV-induced pneumonia was inversely related to the level of maternal neutralizing antibody in infants with a mean age of 4.9 months (Lamprecht et al. 1976). Likewise, serum neutralizing antibody titer has been shown to correlate inversely with the rate of infection (Henderson et al. 1979; Fernald et al. 1983). Finally, the amount of serum IgG to the RSV F protein has correlated with a protective effect against RSV reinfection and illness severity (Kasel et al. 1987). These findings imply that high serum titers of neutralizing antibody should provide at least partial protection against RSV infection and raised the possibility of prevention against serious RSV disease by passive transfer of serum immunoglobulins. In agreement with this, intraperitoneal administration of serum derived from RSV-convalescent cotton rat or human, or of purified human gamma globulin derived from pooled plasma (Sandoglobulin, Novartis Pharmaceuticals, East Hanover, NJ; of note: Sandoglobulin is a general gamma globulin that is not particularly enriched for RSV neutralization), has been shown to prevent RSV replication in the lungs of cotton rats upon subsequent intranasal challenge with the virus (Prince et al. 1985a, b). A serum neutralizing antibody titer of at least 1:380 in the recipient animals conferred almost complete protection against viral replication in the lungs. Additionally, treatment of RSV-infected owl monkeys with intravenously administered Sandoglobulin diminished RSV replication in the lungs of the animals (Hemming et al. 1985). Similar results were achieved with RSV-neutralizing monoclonal antibodies in mice (Taylor et al. 1984) and cotton rats (Walsh et al. 1984). In contrast to the early human vaccination experience, no evidence of enhanced pathology was observed in any of these animal models with administration of anti-RSV antibodies in the presence of RSV.

2   Development of RespiGam (RSV-IGIV)

The successful prevention of RSV replication in the presence of specific neutralizing antibody in animal models without the enhanced pathology in the lungs raised the possibility of passive immunoprophylaxis with IgG in infants to protect against RSV infection. Initial trials used standard intravenous immunoglobulin (IGIV) in high-risk infants and demonstrated the safety of IGIV use in the prevention of RSV disease; the results showed a trend toward lessening the severity of the disease and to shorter hospitalization in the treatment groups compared to the control groups. However, the studies failed to demonstrate unequivocal efficacy (Groothuis et al. 1991; Meissner et al. 1993). The data were interpreted as indicating that the standard IGIV preparations used in these initial trials lacked sufficient titers of RSV-neutralizing antibodies. This then led to the development of an immune globulin preparation that was enriched for RSV-neutralizing antibodies (Siber et al. 1992). A microneutralization assay was used to select for plasma from donors that contained high levels of antibodies that neutralized virus rather than those simply with a high RSV antibody titer; this gave rise to the production of an RSV hyperimmune globulin, RSV-IGIV, which possessed an about fivefold better capacity to neutralize RSV virus than conventional IGIV.

Several multicenter clinical studies were performed to evaluate the prevention of RSV disease by prophylaxis with RSV-IGIV (Groothuis et al. 1993; The PREVENT Study Group 1997; Simoes et al. 1998). Based on the results of these trials, the U.S. Food and Drug Administration (FDA) licensed RespiGam (RSV-IGIV; MedImmune, Gaithersburg, MD) on 18 January 1996 for the prevention of serious lower respiratory tract infection caused by RSV in children younger than 24 months of age with bronchopulmonary dysplasia (BPD) or a history of premature birth (≤35 weeks gestation). RespiGam represented an important first step in preventing severe respiratory tract illness due to RSV (of note: RespiGam is no longer commercially available; it has been replaced by another drug, palivizumab, which is discussed in Sect. 3). It is worth noting that both RespiGam and Sandoglobulin (human gamma globulin) had been tested separately in the therapeutic treatment of RSV-infected infants or children. However in both of these studies, at the doses used, neither of the antibodies was shown to have a significant therapeutic effect (Rimensberger et al. 1996; Rodriguez et al. 1997).

3 Development of Synagis (Palivizumab)

3.1 Generation and Preclinical Development of Palivizumab

Although RSV-IGIV was effective for immunoprophylaxis against severe RSV disease, it needed to be administered intravenously, typically by infusion over several hours, and in relatively large volumes. In addition, RSV-IGIV is isolated from human blood, which is more complicated than the production of antibodies produced recombinantly. Also, because it is a mixed human IgG product, RSV-IGIV, like other IGIVs, may potentially affect subsequent vaccination schedules of certain vaccines, such as for measles, mumps, and rubella (American Academy of Pediatrics 1997).

For the reasons outlined above, we subsequently developed a potent anti-RSV prophylactic agent Synagis (palivizumab; MedImmune, Gaithersburg, MD), which is a humanized mAb (IgG1/κ) that binds to RSV F protein. The molecule was constructed by grafting the six complementarity-determining regions (CDRs) of a murine monoclonal antibody (mAb 1129) to human frameworks (Johnson et al. 1997). mAb 1129 was derived from BALB/c mice that were initially immunized by intranasal infection with the A2 strain of RSV. This initial immunization was followed by successive intraperitoneal inoculations of recombinant F protein and intravenous infusion of purified A2 RSV (Beeler and Coelingh 1989). The splenic lymphocytes of these mice were fused to a murine myeloma cell line to generate the 1129 hybridoma, which secreted an antibody that neutralized a broad spectrum of RSV isolates. The corresponding light and heavy chain variable domain genes from the mouse 1129 hybridoma were cloned and sequenced. The three CDRs of mAb 1129 light chain were transplanted onto the first three frameworks of human light chain K102 (Kabat et al. 1991). The fourth framework was derived from the human light chain gene Jκ4 (Hieter et al. 1982). Likewise, the three CDRs of mAb 1129 heavy chain were transplanted onto the first framework of human heavy chain Cor (Kabat et al. 1991) and the second, third, and fourth frameworks of human heavy chain CE-1 (Kabat et al. 1991). Palivizumab (also initially referred to as MEDI-493) was generated by the cloning of this humanized Fv in frame with the constant regions of a human IgG1/κ; palivizumab heavy and light chain V region frameworks are over 98% human. Of note, Johnson et al. (1997) also reported a frame-shift that resulted in the substitution of the first four residues of the light chain CDR1 (LCDR1), SASS, by four random, nonhuman/nonmouse residues, KCQL.

Palivizumab had a binding affinity (K_d, ~1–2 nM) to the RSV F protein that was very similar to that of its chimeric counterpart and showed a high specific activity and potency in the ability to neutralize various subtypes of RSV in vitro and in vivo. The in vitro assays demonstrated that palivizumab neutralized the RSV subtypes A and B with an EC_{50} of 0.1 and 0.17 µg/ml using assays for microneutralization and fusion inhibition, respectively. In these assays, palivizumab showed an approx. 20- to 30-fold greater potency over RSV-IGIV (Johnson et al. 1997).

The in vivo efficacy studies were designed to assess the prophylactic activity of palivizumab in cotton rats. These challenge studies indicated that palivizumab was effective in reducing the replication of RSV subtypes A and B in the lungs by more than 99% when injected intramuscularly at a dose of 2.5 mg/kg. This corresponded to serum concentrations of about 30 µg/ml. Treatment with palivizumab did not enhance RSV-induced pathology or RSV replication during primary infection (Porterfield 1986), nor did it affect the development of a protective anti-RSV immune response or result in increased pulmonary pathology upon a secondary challenge (Kapikian et al. 1969). In light of these promising preclinical results, we selected palivizumab as a lead candidate for RSV prophylaxis in high-risk human infants.

3.2 Clinical Development of Palivizumab and Its Licensure

Several multicenter clinical studies were performed to evaluate the prevention of RSV disease by prophylaxis with palivizumab (Subramanian et al. 1998; Saez-Llorens et al. 1998; The IMpact-RSV Study Group 1998). In the IMpact study (1998), over 1,500 premature infants and young children less than 2 years old with BPD were recruited and treated with either five monthly intramuscular injections of palivizumab at 15 mg/kg or an equivalent volume of placebo. In the prophylactic palivizumab arm an overall 55% reduction in RSV-related hospitalization incidence ($p < 0.001$) was found over that in the placebo arm. Based on the results of these multiple trials, in 1998 the FDA approved Synagis (palivizumab) administered via monthly intramuscular injection for the immunoprophylaxis of serious RSV respiratory disease in high-risk infants and children, such as those with BPD or with a history of premature birth. Subsequent to FDA licensure of palivizumab, a placebo-controlled trial of palivizumab in young children with congenital heart disease (CHD) ($n=1287$) was conducted (Feltes et al. 2003). In the palivizumab treatment group a 45% relative reduction in RSV hospitalizations ($p=0.003$) over that in the control group was observed. The data from this study showed that monthly palivizumab administration (15 mg/kg intramuscularly) was safe, well-tolerated, and effective for prophylaxis of serious RSV disease in young children with hemodynamically significant CHD. This study was the pivotal clinical trial for FDA approval for use in this patient population.

Palivizumab was also approved in Europe in 1999. Subsequently, bridging studies to evaluate the safety and efficacy of palivizumab in the Japanese population were also carried out (Groothuis and Nishida 2002), and the results led to its approval in Japan in 2002. Of equal importance, the consistency and robustness of the palivizumab manufacturing process in terms of cell line stability, varying scales of production, and location of manufacturing sites was also established (Schenerman et al. 1999).

In summary, palivizumab is indicated for the prevention of serious lower respiratory tract disease caused by RSV in pediatric patients at high risk of RSV disease. These include infants with BPD and infants with a history of premature birth, as well as children with hemodynamically significant CHD. The recommended dose of palivizumab is 15 mg/kg. The first dose should be administered prior to commencement of the RSV season. In the northern hemisphere, the RSV season typically begins in November and lasts through April, but it may begin earlier or persist later in certain communities [Synagis (palivizumab), Package insert, MedImmune, Inc.; product information as of 10 March 2006].

3.3 Ability of Palivizumab to Neutralize Clinical RSV Isolates

The ability of palivizumab to bind to and neutralize a broad spectrum of RSV subtypes is likely linked to its overall clinical efficacy. In one study, palivizumab bound to over 700 RSV isolates that originated from 19 countries (Branco et al. 2000). Palivizumab also bound to and/or neutralized another 47 subtype A and 33 subtype B RSV isolates, including samples from Japan (Johnson et al. 1997; Groothuis and Nishida 2002). Finally, palivizumab was able to bind to 100% of a panel of 371 RSV isolates from infants that were hospitalized for RSV disease in the United States (DeVincenzo et al. 2004). The samples were collected over the first 4 years following the licensure of palivizumab in 1998 and included 25 "breakthrough" recipients (i.e., from patients that received palivizumab but still had RSV infections). Taken together, these data have established the broad reactivity of palivizumab. RSV mutants resistant to palivizumab or its parental murine counterpart have been generated in culture or in immunosuppressed cotton rats (Beeler and Coelingh 1989; Crowe et al. 1998a; Zhao et al. 2004; Zhao and Sullender 2005), and palivizumab likely exerts a selective pressure on RSV in treated infants (DeVincenzo et al. 2003, 2004). However, in studies to date, in the patients that received mAb but were hospitalized due to breakthrough RSV disease in mAb, there does not appear to be a rapid selection of viral mutants that are resistant to the mAb, and the monitoring of clinical isolates has not detected the appearance of palivizumab-resistant viral strains in the human population. It is possible that the decreased fitness of some of these variants would preclude their dissemination in the population (Zhao et al. 2006). However, at present, the clinical significance and relevance of mAb-resistant variants remain largely unknown.

3.4 The Recognition Site of Palivizumab

The F glycoprotein is one of the two major surface proteins of RSV. This homotrimeric protein, initially synthesized as an inactive precursor (F_0), is cleaved during its maturation process into two disulfide-linked chains (F_1 and F_2; Collins and Mottet 1991). Once activated by cleavage, the protein allows the viral and infected cell membranes to fuse, and thus it promotes the formation of syncytia (Walsh and Hruska 1983). The F protein constitutes a prime antibody target for neutralizing viral infectivity (Merz et al. 1980; Murphy and Walsh 1988) because of its high degree of conservation both between and within RSV subtypes (Johnson et al. 1987; Lopez et al. 1988).

Several techniques have enabled the mapping of neutralizing epitopes on the surface of the RSV F protein. For instance, competition between the various neutralizing antibodies for binding to RSV has enabled the broad definition of distinct antigenic sites, and this has led to the recognition of at least three major antigenic areas on its surface (A, B, and C; Beeler and Coelingh 1989). Further definition of specific binding sites has come from studying the interaction of the antibodies with F-derived peptides or protein fragments (Trudel et al. 1987; Lopez et al. 1990; Bourgeois et al. 1991; Martin-Gallardo et al. 1991). Another approach to map the epitopes has come from analyses of the sequence changes in the virus following the generation of mAb-resistant mutants of RSV (MARMs) in cell culture. The amino acid sequences of the mutants have revealed sequence changes in the protein that regulate antibody binding (Beeler and Coelingh 1989; Garcia-Barreno et al. 1989; Arbiza et al. 1992; Lopez et al. 1998). Combined, these studies have identified at least five distinct neutralizing antigenic sites on the surface of the F protein (I, II, IV, V, and VI), as shown in Fig. 1. Attempts to correlate these sites to the three previously defined antigenic areas revealed that the antigenic areas A, B, and C comprised components of antigenic sites II, I, and IV/V/VI, respectively.

Much of our current knowledge regarding the epitope of palivizumab and its parental murine counterpart has been derived from competitive binding analysis of mAbs to RSV and through the generation of MARMs. These approaches indicated that at least part of the epitope bound by palivizumab mapped in the F_1 subunit of the RSV F glycoprotein to the antigenic area A (Beeler and Coelingh 1989) and involved residues at positions 272 and 275, using the amino acid numbering of the full-length F_0 form (Crowe et al. 1998a; Zhao et al. 2004; Zhao and Sullender 2005). These

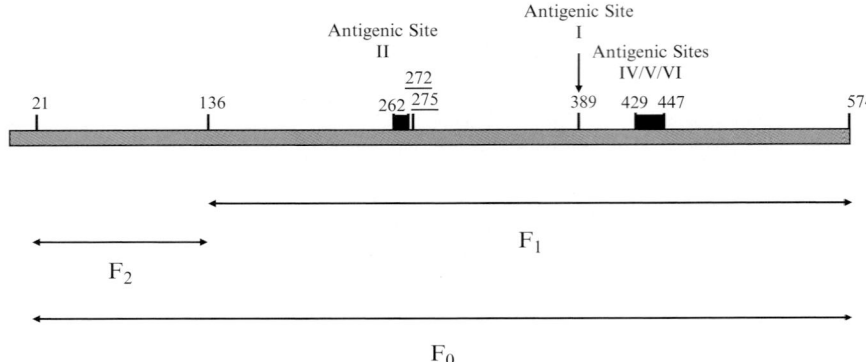

Fig. 1 Location of neutralizing epitopes within the primary structure of the RSV F glycoprotein. Amino acid numbers are for the full-length F_0 form. Residues 1–20, 21–136 and 137–574 correspond to the signal peptide, F_2 and F_1 subunits, respectively (according to GeneBank entry BAA00240). Antigenic sites are centered around (*vertical arrow*), or spanning (*boxes in black*) the positions at which mutations are known to occur in the corresponding MARMs. Residues critical for the interaction of the RSV F protein with palivizumab are *underlined*

residues corresponded to antigenic site II (Crowe et al. 1998b). The spatial location of the antigenic area recognized by palivizumab can be broadly defined, even though no X-ray or nuclear magnetic resonance (NMR)-derived three-dimensional structure of the RSV F protein has been reported. A three-dimensional model of RSV F protein exists, and this has revealed a cone-shaped structure consisting of three regions, namely head, neck, and stalk (Smith et al. 2002). Based on this model, the epitope of palivizumab is located on the exterior of the neck of the F protein. This finding is in good agreement with studies of the complex of RSV F protein and various monoclonal antibodies by electron microscopy, which showed that site II is located near the base of the F protein head (Calder et al. 2000).

However, despite the body of knowledge regarding the epitope of palivizumab, the molecular mechanism by which this antibody neutralizes RSV has yet to be elucidated. Various hypotheses have been suggested, which include the inhibition of the binding of the F protein fusion domain to the infected cell or the prevention of the structural rearrangements of the F protein required prior to fusion (Johnson et al. 1999).

3.5   Comparison of Palivizumab with Other Monoclonal Antibodies

Over the years several other anti-RSV monoclonal antibodies have been developed to prevent RSV infections. Among these, RSHZ19 (also referred to as SB 209763), is a humanized IgG1/κ developed by Glaxosmithkline. The epitope of RSHZ19 is distinct from, though functionally overlapping with, that of palivizumab (Johnson et al. 1999). RSHZ19 exhibited both a curative and prophylactic effect in RSV-infected mice by binding to a neutralizing epitope on the RSV F protein (Tempest et al. 1991). However, despite these promising early results and an established safety profile in human (Everitt et al. 1996), in a clinical phase III study RSHZ19 failed to significantly protect high-risk infants against RSV infection. A direct comparison has indicated that the lack of efficacy of RSHZ19 in humans likely was due to an overall 2- to 5-fold lower potency than palivizumab in the inhibition of RSV replication in vivo, the binding affinity to RSV F protein, and microneutralization and inhibition of fusion (Johnson et al. 1999).

HNK20 is a mouse monoclonal IgA that is directed against the RSV F glycoprotein that was developed by OraVax. Administration of the mAb to the upper airway of monkeys resulted in high titers of the neutralizing IgA in both serum and nasal secretions, and the mAb also significantly reduced virus shedding (Weltzin et al. 1996). However, in a clinical phase III trial no significant prophylactic effectiveness was seen in humans, and development of HNK20 was apparently discontinued.

The use of phage display technology has led to the identification of fully human Fab fragments that are directed against neutralizing epitopes on the RSV F glycoprotein (Barbas et al. 1992). One such Fab (Fab19) showed potent neutralization of RSV in vitro. Upon intranasal administration to RSV-infected mice, the Fab potently neutralized the virus and decreased lung RSV titers (Crowe et al. 1994). The authors of this work suggested that this Fab might be used for the short-term prophylaxis of RSV disease in high-risk individuals via direct nasal administration. Another Fab (Fab RSVF2–5) exhibited broad activity in neutralizing RSV subtypes A and B and also showed significant therapeutic efficacy upon intranasal instillation on RSV-infected mice (Crowe et al. 1998b). However, none of these antibodies or antibody fragments has yet progressed into clinical testing.

In summary, despite a significant interest and active efforts in the targeting of RSV in both biotechnological and academic settings, palivizumab remains the only monoclonal antibody approved for the prevention of serious RSV-related lower respiratory tract infection in high-risk infants and young children.

4   Development of Motavizumab

The successful development of RSV-IGIV and palivizumab has validated the approach of immunoprophylaxis for the prevention of RSV infection. The improved potency of palivizumab over RespiGam has resulted in a dramatic reduction in monthly dosing of IgG, from 750 mg/kg to 15 mg/kg, while maintaining similar clinical efficacy. The lower dose made it possible to formulate the drug (at 100 mg/ml) for intramuscular administration. Although RespiGam and palivizumab have successfully reduced RSV hospitalizations for infants and young children at high risk (as discussed in earlier sections), a question remains whether this clinical efficacy could be improved. One way to better protect these patients would be to increase the prophylactic dose; however, this likely would require a formulation concentration much higher than 100 mg/ml (for intramuscular administration), which has not been easy to achieve. Alternatively, an antibody with higher potency than palivizumab might protect more patients from RSV infection; we have adopted this approach.

Following the licensure of palivizumab in 1998, several strategies were undertaken to develop a more effective neutralizing mAb against RSV. Cutting-edge technologies in the antibody field have been employed; these include phage display, hybridomas from transgenic mice expressing human antibodies, and protein engineering by directed evolution. Though human antibodies against the RSV F protein have been successfully identified through the first two approaches, none of the resulting mAbs has shown an enhanced ability to neutralize virus over that seen with palivizumab. In contrast, as outlined below, engineering of palivizumab has been successful in improving the activity of the mAb. In spite of several unexpected outcomes that had to be addressed, the final candidate, motavizumab, has improved potency over palivizumab both in vitro and in vivo.

Our first step in engineering palivizumab was to restore the parental LCDR1 sequence. In previous studies, due to a synthetic error in the humanization process, four random, nonhuman, nonmouse residues (KCQL) were introduced in LCDR1 instead of the original SASS residues (Johnson et al. 1997). Although these residues did not appear to impact the affinity significantly and the free cysteine did not pose problems with antibody production, we decided to revert these to the original sequence. In addition, we mutated two murine residues in the palivizumab sequence to human residues (A105Q on the heavy chain and L104V on the light chain). Previously, the murine residues were retained based on the molecular modeling that implicated their potential involvement in F protein binding. However, after the changes were made, we did not observe any significant changes in antibody affinity, which indicates that A105 and L104 are unlikely to be involved in binding. The engineered antibody, 493L1FR, has a fully human framework sequence and similar binding affinity to F protein as palivizumab. It was used as a starting template for subsequent affinity maturation engineering (Wu et al. 2005). All residues discussed in this article are numbered according to the Kabat numbering system (Kabat et al. 1991).

4.1   Improvement of Dissociation Rate and Its Impact on Viral Neutralization

In nature, high-affinity antibodies are generated by multiple rounds of somatic hypermutation followed by preferential B cell clonal selection. We have developed a directed evolution approach that mimics this natural process. Point mutations were generated in the CDR regions by an approach of oligonucleotide-based mutagenesis. Variants in the Fab format were expressed in *Escherichia coli* and an ELISA-based screening procedure was developed to select for beneficial mutations that enhanced the affinity. Combinatorial libraries that included the beneficial point mutations were constructed and then subjected to another round of selection (Wu et al. 1998). In a period of about 3 months, many beneficial point mutations were identified and several high-affinity combinatorial variants were selected (Wu et al. 2005). These point mutations are located at four CDR positions: S32A and S32P at heavy chain CDR1 (HCDR1), W100F at heavy chain CDR3 (HCDR3), S52F and S52Y at light chain CDR2 (LCDR2), and G93F, G93Y, and G93W at light chain CDR3 (LCDR3). Affinity measurements of these variants in Fab format by the BIAcore biosensor showed a three- to sevenfold increase in binding to F protein when compared with 493L1FR Fab. The best combinatorial Fab variants (K_d, 37 to 45 pM) contained three or four beneficial mutations and exhibited a ≥117-fold higher affinity than the palivizumab Fab (K_d, 5.25 nM). The affinity improvement observed in these variants (both single-mutation and combinatorial variants) is largely driven by a decrease in the dissociation rate (k_{off}).

To characterize the antiviral activities of these k_{off} variants, we used an assay that measures RSV microneutralization (Anderson et al. 1985; Johnson et al. 1997). The neutralization titers (IC_{50}) of the combinatorial Fab variants ranged from 1.43 to 4.98 nM (0.0715 to 0.249 µg/ml) and were 110- to 384-fold more potent than palivizumab Fab (IC_{50}, 549.2 nM or 27.46 µg/ml) (Wu et al. 2005). In vitro we observed an excellent correlation between the affinity and the ability to neutralize virus among the single-mutation and combinatorial Fab variants. However, upon conversion of these Fab variants into IgGs we encountered unexpected results. We observed little to no improvement in the ability of these IgGs to neutralize RSV over that observed for palivizumab IgG. The IC_{50} values of the variants ranged from 2.04 to 3.47 nM (0.306 to 0.521 µg/ml) versus that of 3.02 nM (0.453 µg/ml) for palivizumab IgG (Wu et al. 2005). Our results showed that the palivizumab, when converted from the Fab to IgG format, had a 182-fold increase in in vitro potency (IC_{50} decreased from 549.2 nM to 3.02 nM). In contrast, when the palivizumab-derived variants were converted to IgG, there was no enhancement in potency. It is not clear why the palivizumab in vitro potency dramatically increases upon conversion from Fab to IgG format, yet that of its k_{off} variants do not. These results cannot be explained by an avidity effect that arises from bivalent IgG binding since the K_d (and k_{off}) of palivizumab improved only very marginally upon conversion to IgG. In addition, the combinatorial variant AFFF IgG, while having an approx. 2 log higher avidity than palivizumab, only showed a minimal increase in potency. To decipher this, a clear understanding of the mechanism at the molecular level for RSV neutralization by palivizumab (and its variants) is necessary.

4.2   Improvement of Association Rate and Its Impact on Viral Neutralization

Although studies have shown that k_{off}-driven affinity maturation improves biological function (reviewed in Chowdhury and Wu 2005), this strategy did not work for our anti-RSV mAb. We decided to explore an alternative strategy by which we would engineer association rate (k_{on}) and study its influence on RSV neutralization. For this purpose we developed a novel k_{on} ELISA screen (Wu and An 2003; Wu et al. 2005).

To favor high k_{on} selection, the interaction time between antibody and antigen was reduced to 10 min. In addition, the number of washes and washing time were reduced to minimize the impact of antibody–antigen complex dissociation (3 washes in <30 s). The best k_{off} variant based on viral neutralization, AFFF, was chosen as the starting template. AFFF contains four beneficial k_{off} mutations located at four CDRs (S32A in HCDR1, W100F in HCDR3, S52F in LCDR2, and G93F in LCDR3) (Fig. 2). We applied an iterative mutagenesis and screening strategy to gradually increase the k_{on}. Altogether, about ten CDR mutation libraries including single, double, and combinatorial mutations were constructed and screened. In contrast to the k_{off} mutations that improved the off rate 3.7- to 7.8-fold, we found that each k_{on} mutation only produced about a 20%–80% increase in the on rate.

Beneficial single k_{on} mutations were identified across all CDR loops, and the best combinatorial variants typically contained about 7–13 k_{on} mutations. These combinatorial variants in the Fab format exhibited an approx. 4- to 5-fold increase in k_{on} and 2- to 13-fold improvement in k_{off} over palivizumab Fab (Wu et al. 2005). Although the starting clone AFFF had an approx. 2 $\log_{10}$ improvement in k_{off}, during the process some of the beneficial k_{off} mutations ended up being replaced with mutations that altered k_{on}. We subsequently evaluated these k_{on} Fab variants in an RSV microneutralization assay and found that they had a vastly improved ability to neutralize virus and the activity observed in these k_{on} variants was substantially greater than that of k_{off} variants. For example, the best k_{off} variant AFFF Fab has a 384-fold improvement while the k_{on} variants A17d4, A12a6, and A13c4 (see Wu et al. 2005 for sequences) have a 1,000- to 1,534-fold improvement. Most importantly, unlike the k_{off} variants, when these k_{on} variants were converted from Fab to IgG they retained much higher potency than that of palivizumab IgG. In addition, unlike palivizumab, some, but not all, of the converted k_{on} variants showed a 3- to 13-fold improvement in k_{off} (over their Fab counterpart) due to the increased avidity of the IgG over the Fab. Following these rounds of this genetic manipulation, the best full-length antibody in terms of ability to neutralize virus is A4b4 (Fig. 2). A4b4 IgG has a 27 pM avidity to RSV F protein, which is 125-fold better than palivizumab. It neutralizes RSV in tissue culture with an IC_{50} of 69 pM; this represents a 44-fold improvement in potency over palivizumab. Our findings also indicate that k_{on} of a mAb can play a dominant role in the neutralization of RSV, particularly when the antibody is in an IgG format. This may be explained in part by the likelihood that higher k_{on} antibodies bind to the virus more quickly and thus neutralize it before it has a chance to infect cells.

4.3 Characterization of Palivizumab Variants: Viral Neutralization, Pharmacokinetics, and Biodistribution

Several of the top k_{on} variants that had a 12- to 44-fold improvement in the ability to neutralize RSV in vitro were further evaluated in an immunoprophylaxis model in cotton rats (Wu et al. 2007). To our surprise, we did not observe the anticipated efficacy improvement to the extent predicted by the in vitro potency of the mAb. In a dose titration study even the best variant, A4b4, achieved only a roughly twofold improvement

VH Domain

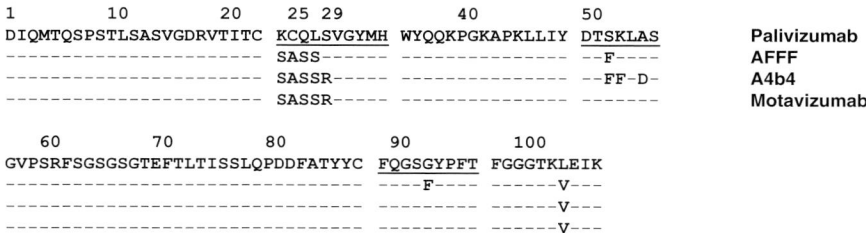

```
1         10        20        30    ab    40
QVTLRESGPALVKPTQTLTLTCTFSGFSLS TSGMSVG WIRQPPGKALEWLA        Palivizumab
------------------------------ -A----- --------------        AFFF
------------------------------ -A----- --------------        A4b4
------------------------------ -A----- --------------        Motavizumab

50        60        70        80 abc      90        100       110
DIWWDDKKDYNPSLKS RLTISKDTSKNQVVLKVTNMDPADTATYYCAR SMITNWYFDV WGAGTTVTSS
--------------- ------------------------------- -----F---- --Q--------
--------H------D ------------------------------- D--F-F---- --Q--------
--------H------D ------------------------------- D--F-F---- --Q--------
```

VL Domain

```
1         10        20    25 29        40        50
DIQMTQSPSTLSASVGDRVTITC KCQLSVGYMH WYQQKPGKAPKLLIY DTSKLAS    Palivizumab
---------------------- SASS------ --------------- --F----    AFFF
---------------------- SASSR----- --------------- --FF-D-    A4b4
---------------------- SASSR----- --------------- -------    Motavizumab

  60        70        80        90        100
GVPSRFSGSGSGTEFTLTISSLQPDDFATYYC FQGSGYPFT FGGGTKLEIK
------------------------------- ----F---- ------V---
------------------------------- --------- ------V---
------------------------------- --------- ------V---
```

Fig. 2 Amino acid sequence alignment of the variable regions of palivizumab, AFFF (k_{off} variant), A4b4 (k_{on} variant), and motavizumab. Residues are numbered according to the Kabat numbering system (Kabat et al. 1991). CDR regions as defined by Kabat are *underlined*

in dose efficacy over palivizumab. When both mAbs were given at the same dose, A4b4 reduced lung viral titers about 3–5 times more effectively than palivizumab. A pharmacokinetic study of A4b4 in cotton rats showed that it had rather poor properties. In a comparative study that included A4b4, AFFF, and palivizumab, 6 days after intramuscular administration, the serum levels of A4b4 had dropped to levels 5-fold lower than those of palivizumab, and after 10 days these were 50-fold lower (Wu et al. 2007). In contrast, AFFF behaved very similarly to palivizumab. We also compared the lung biodistribution of these antibodies. Again A4b4 was different and the levels were substantially lower in lung homogenates and bronchial alveolar lavage (BAL) fluid than those of palivizumab or AFFF. A4b4 was further tested in mice and cynomolgus monkeys, and again poor pharmacokinetic properties and lower distribution into the lung were observed. The poor pharmacokinetic and bioavailability properties of A4b4 probably explain the initial apparent discrepancy between the observed potency in vitro and results for ability to neutralize virus in vivo.

To explore the reasons for the poor in vivo properties of A4b4, we investigated the stability of A4b4 in circulation, the immunogenicity against A4b4 in cotton rats, and also looked for nonspecific binding of the mAb to various tissues. Our studies ruled out the first two causes; however, our immunohistochemical analyses revealed that A4b4 stained a broad array of tissues, including lung, liver, spleen, skeletal muscle, intestine, and skin from normal human and cotton rat, and lung tissues from cynomolgus monkey and chimpanzee (Wu et al. 2007). Such broad cross-reactivity may result in the rapid removal of A4b4 from blood circulation and may also cause the low lung bioavailability. Such cross-reactive binding has not been observed with palivizumab.

4.4   Improvement of the in vivo Properties of Palivizumab Variant A4b4

Since A4b4 was built upon AFFF through the engineering of k_{on}, and AFFF was found to have pharmacokinetic and biodistribution properties that were similar to palivizumab, it was likely that some of the k_{on} mutations (in the context of the remaining k_{off} mutations) in A4b4 contributed to the broad tissue binding, and hence negatively affected the pharmacokinetic and biodistribution properties of the mAb. Since these k_{on} mutations were distributed among the heavy and light chain CDRs, we generated chain-swapped variants in which we paired the palivizumab heavy chain (HC) with the A4b4 light chain (LC), and palivizumab LC with A4b4 HC. The lung bioavailability of both variants together with A4b4 and palivizumab was assessed in cotton rats. Only the antibodies that contained the A4b4 LC had poor distribution to the lung, thus indicating that mutations on A4b4 LC played a role in this poor biodistribution. Subsequently, we engaged in a thorough mutagenesis study of these mutations (located at LC positions 29, 52, 53, and 55) for their impact on the in vivo properties (Wu et al. 2007). Point mutations were generated to revert each position respectively to the palivizumab sequence. A4b4 LCs containing each of these reverse mutations were coexpressed with the A4b4 HC, and the resultant antibodies were tested for their ability to neutralize RSV in tissue culture and were also assessed for their lung bioavailability in cotton rats. In addition, individual mutations were made on the 493L1FR LC to the A4b4 residue (S29R and S52F), and these mutated 493L1FR LCs were coexpressed with the A4b4 HC. The resultant molecules were also tested both in vitro and in vivo. From these studies, we found two molecules that displayed properties for lung biodistribution that were similar to those of palivizumab; these molecules were also very potent in viral neutralization. These two leads were tested further for their pharmacokinetic and biodistribution properties in cynomolgus monkeys. Only one of the leads exhibited properties that were comparable to palivizumab; it comprised the A4b4 HC and a mutated 493L1FR LC containing a S29R forward mutation. This clone, motavizumab, was selected as the final clinical candidate.

4.5   Motavizumab: A Second-Generation Anti-RSV mAb

Motavizumab (also known as MEDI-524) contains six CDR changes and one framework change in the HC and five CDR changes and one framework change in the LC compared to its parent palivizumab (Fig. 2). The binding affinity of motavizumab (K_d, 34.6 pM) for RSV F protein is about 70-fold greater than palivizumab, and it has a 6-fold faster k_{on} and an 11-fold slower k_{off} than its parent (Wu et al. 2007). Motavizumab is about 20 times more potent than palivizumab in the neutralization of RSV in tissue culture. Studies that followed the ability of the mAbs to protect cotton rats from RSV infection showed that, at equivalent doses, motavizumab reduced RSV titers in the lung 10–100 times more than palivizumab (depending on the RSV strain). Furthermore, at a dose of 2 mg/kg, motavizumab reduced the RSV titers in nasal turbinates by 2.8 $\log_{10}$ while palivizumab reduced titers by only 0.7 $\log_{10}$. These results indicate that this new anti-RSV antibody likely will give rise to improved viral neutralization in both the upper and lower respiratory tracts in humans. Motavizumab will thus potentially provide an efficacy superior to that of palivizumab in the prevention of severe RSV illness in high-risk infants and children. Such efficacy studies are currently being carried out in pivotal clinical trials. The results from the phase I and II trials have shown that the drug is safe. Moreover, in a preliminary analysis of data from the comparative phase III pivotal study (data not shown), motavizumab achieved its primary endpoint of noninferiority by reducing the incidence of hospitalizations caused by RSV in infants at high risk for serious RSV disease by 26% when compared to palivizumab. The data also indicate that motavizumab showed statistical superiority over palivizumab in a secondary endpoint by reducing the incidence of RSV-specific, medically attended outpatient lower respiratory infection by 50%. If these results hold true we will have validated, for the first time, an affinity maturation approach to improve mAb efficacy for human therapy.

5 Long-Lasting Anti-RSV mAb: Numax-YTE

Both palivizumab and motavizumab have a circulation half-life of up to 3 weeks; thus they need or will need to be administered monthly in order to maintain effective IgG levels to prevent RSV infection. To cover the RSV season in any of the geographic locations where the disease is seen, several monthly injections are necessary. Thus to ensure full protection for high-risk infants, full compliance will be needed with monthly visits to the doctors' office.

Recent studies have shown that the neonatal Fc receptor (FcRn) plays a key role in maintaining the serum IgG level (reviewed by Junghans 1997; Ghetie and Ward 2000). This receptor is responsible for the long serum half-life observed with IgG. Binding of IgG to FcRn is pH dependent. It binds tightly at pH 6 and exhibits almost no binding at pH 7.4. This property of pH-dependent binding enables IgG recycling to occur in which IgGs in circulation are taken up into cells by pinocytosis. Following uptake, IgGs that bind to FcRn in the acidic endosomes are recycled to the cell surface and then released back into the circulation. IgGs that do not bind FcRn enter the lysosomal pathway and are degraded by proteases. Based on this mechanism, one can hypothesize that an engineered IgG with a higher affinity to FcRn at pH 6 than wild-type IgG, but still with no detectable binding at neutral pH, should compete favorably with wild-type IgG for FcRn binding and hence have a longer half-life (better recycling).

To improve the half-life of our therapeutic antibodies, we have engaged in engineering the human Fc region for improved affinity to human FcRn at pH 6. Through rationally designed libraries and phage display selection, we have identified variants with up to 57-fold increases in affinity (Dall'Acqua et al. 2002). One of the Fc variants has been further characterized as a variant of motavizumab. The Fc changes consisted of a triple substitution, M252Y/S254T/T256E (EU numbering; Kabat et al. 1991), in the CH2 region of motavizumab. The engineered antibody, named Numax-YTE (alternatively MEDI-524-YTE or MEDI-557), has approx. a tenfold higher affinity to human FcRn at pH 6 compared to motavizumab, but has almost no detectable binding to human FcRn at pH 7.4. It shows similar binding properties to cynomolgus monkey FcRn. The pharmacokinetic and biodistribution properties of this molecule were tested in cynomolgus monkeys (Dall'Acqua et al. 2006). In these animals, Numax-YTE consistently showed a fourfold increase in serum half-life compared with the parental motavizumab. Furthermore, the sustained serum IgG levels resulted in a fourfold increase in lung bioavailability of the Numax-YTE. We plan to test Numax-YTE in humans to determine whether this favorable pharmacokinetic profile is retained. If successful, Numax-YTE may offer the opportunity for less frequent dosing. It may also be possible to develop a long-lasting anti-RSV mAb that requires administration only once or twice per season for full protection, through a combination of increased half-life and further potency improvement, together with a higher dose of mAb. This may improve the prevention of RSV by improving both efficacy and compliance. Importantly, if this long-half-life YTE technology is validated in humans, it can be applied broadly to many other therapeutic antibodies and will likely revolutionize the field of antibody therapy.

6   Conclusions

In the past two decades, there has been immense growth in novel antibody technologies, enabling researchers to readily generate highly specific and high-affinity antibodies against almost any disease target and to further manipulate antibody properties beyond their natural limitations. In this report we summarize our efforts in developing anti-RSV antibodies using the most advanced tools available. In our study, through mutagenesis of the mAb sequence, we were able to engineer both the in vitro and in vivo properties of the antibody, including affinity, tissue cross-reactivity, and serum half-life. We have found that there is a major differential effect on the neutralization of RSV caused by the on and off rate of the mAb, with the on rate playing a dominant role. In addition, our engineering of the pH-specific binding properties of the antibody Fc domain to FcRn has led to an increased antibody serum half-life by a few fold in a primate model. Its application to RSV intervention may lead to the development of a vaccine-like immunoprophylactic agent in terms of its potential prolonged effect of protection. Similarly, this half-life extension technology can be used to generate long-lasting antibody therapeutics for other diseases.

References

American Academy of Pediatrics (1997) Measles. In: Peter G (ed) Red book. Report of the committee on infectious diseases, 24th edn. American Academy of Pediatrics, Elk Grove Village, p 353
Anderson LJ, Hierholzer JC, Bingham PG, Stone YO (1985) Microneutralization test for respiratory syncytial virus based on an enzyme immunoassay. J Clin Microbiol 22:1050–1052
Arbiza J, Taylor G, Lopez JA, Furze J, Wyld S, Whyte P, Stott EJ, Wertz G, Sullender W, Trudel M, Melero JA (1992) Characterization of two antigenic sites recognized by neutralizing monoclonal antibodies directed against the fusion glycoprotein of human respiratory syncytial virus. J Gen Virol 73:2225–2234
Barbas CF 3rd, Crowe JE Jr, Cababa D, Jones TM, Zebedee SL, Murphy BR, Chanock RM, Burton DR (1992) Human monoclonal Fab fragments derived from a combinatorial library bind to respiratory syncytial virus F glycoprotein and neutralize infectivity. Proc Natl Acad Sci USA 89:10164–10168
Beeler JA, van Wyke Coelingh K (1989) Neutralization epitopes of the F glycoprotein of respiratory syncytial virus: effect of mutation upon fusion function. J Virol 63:2941–2950
Blount RE, Morris JA, Savage RE (1956) Recovery of cytopathogenic agent from chimpanzees with coryza. Proc Soc Exp Biol Med 92:544–549
Bourgeois C, Corvaisier C, Bour JB, Kohli E, Pothier P (1991) Use of synthetic peptides to locate neutralizing antigenic domains on the fusion protein of respiratory syncytial virus. J Gen Virol 72:1051–1058
Branco L, Johnson LS, Young JF, Tressler R, McCue M, Groothuis JR (2000) Broad range binding of palivizumab to respiratory syncytial virus (RSV) [abstr]. 17th European Congress of Perinatal Medicine. Congress Society, p 43
Calder LJ, Gonzalez-Reyes L, Garcia-Barreno B, Wharton SA, Skehel JJ, Wiley DC, Melero JA (2000) Electron microscopy of the human respiratory syncytial virus fusion protein and complexes that it forms with monoclonal antibodies. Virology 271:122–131
Chowdhury PS, Wu H (2005) Tailor-made antibody therapeutics. Methods 36:11–24
Collins PL, Mottet G (1991) Post-translational processing and oligomerization of the fusion glycoprotein of human respiratory syncytial virus. J Gen Virol 72:3095–3101
Collins PL, Chanock RM, Murphy BR (2001) Respiratory syncytial virus. In: Knipe DM, Howley PM, Griffin DE, Lamb RA, Martin MA, Roizman B, Straus SE (eds) Fields virology, 4th edn. Lippincott Williams & Wilkins, Philadelphia, pp 1443–1485
Crowe JE, Firestone CY, Crim R, Beeler JA, Coelingh KL, Barbas CF, Burton DR, Chanock RM, Murphy BR (1998a) Monoclonal antibody-resistant mutants selected with a respiratory syncytial virus-neutralizing human antibody fab fragment (Fab 19) define a unique epitope on the fusion (F) glycoprotein. Virology 252:373–375
Crowe JE Jr (2001) Influence of maternal antibodies on neonatal immunization against respiratory viruses. Clin Infect Dis 33:1720–1727
Crowe JE Jr, Murphy BR, Chanock RM, Williamson RA, Barbas CF 3rd, Burton DR (1994) Recombinant human respiratory syncytial virus (RSV) monoclonal antibody Fab is effective therapeutically when introduced directly into the lungs of RSV-infected mice. Proc Natl Acad Sci USA 91:1386–1390
Crowe JE Jr, Gilmour PS, Murphy BR, Chanock RM, Duan L, Pomerantz RJ, Pilkington GR (1998b) Isolation of a second recombinant human respiratory syncytial virus monoclonal antibody fragment (Fab RSVF2–5) that exhibits therapeutic efficacy in vivo. J Infect Dis 177:1073–1076
Dall'Acqua WF, Woods RM, Ward ES, Palaszynski SR, Patel NK, Brewah YA, Wu H, Kiener PA, Langermann S (2002) Increasing the affinity of a human IgG1 for the neonatal Fc receptor: biological consequences. J Immunol 169:5171–5180
Dall'Acqua WF, Kiener PA, Wu H (2006) Properties of human IgG1s engineered for enhanced binding to the neonatal Fc receptor (FcRn). J Biol Chem 281:23514–23524

DeVincenzo JP, Aitken J, Harrison L (2003) Respiratory syncytial virus (RSV) loads in premature infants with and without prophylactic RSV fusion protein monoclonal antibody. J Pediatr 143:123–126

DeVincenzo JP, Hall CB, Kimberlin DW, Sanchez PJ, Rodriguez WJ, Jantausch BA, Corey L, Kahn JS, Englund JA, Suzich JA, Palmer-Hill FJ, Branco L, Johnson S, Patel NK, Piazza FM (2004) Surveillance of clinical isolates of respiratory syncytial virus for palivizumab (Synagis)-resistant mutants. J Infect Dis 190:975–978

Everitt DE, Davis CB, Thompson K, DiCicco R, Ilson B, Demuth SG, Herzyk DJ, Jorkasky DK (1996) The pharmacokinetics, antigenicity, and fusion-inhibition activity of RSHZ19, a humanized monoclonal antibody to respiratory syncytial virus, in healthy volunteers. J Infect Dis 174:463–469

Falsey AR, Walsh EE (2000) Respiratory syncytial virus infection in adults. Clin Microbiol Rev 13 :371–384

Feltes TF, Cabalka AK, Meissner HC, Piazza FM, Carlin DA, Top FH Jr, Connor EM, Sondheimer HM, The CARDIAC Synagis Study Group (2003) Palivizumab prophylaxis reduces hospitalization due to respiratory syncytial virus in young children with hemodynamically significant congenital heart disease. J Pediatr 143:532–540

Fernald GW, Almond JR, Henderson FW (1983) Cellular and humoral immunity in recurrent respiratory syncytial virus infections. Pediatr Res 17:753–758

Garcia-Barreno B, Palomo C, Penas C, Delgado T, Perez-Brena P, Melero JA (1989) Marked differences in the antigenic structure of human respiratory syncytial virus F and G glycoproteins. J Virol 63:925–932

Garvie DG, Gray J (1980) Outbreak of respiratory syncytial virus infection in the elderly. Br Med J 281:1253–1254

Ghetie V, Ward ES (2000) Multiple roles for the major histocompatibility complex class I-related receptor FcRn. Annu Rev Immunol 18:739–766

Glezen WP, Taber LH, Frank A, Kasel JA (1986) Risk of primary infection and reinfection with respiratory syncytial virus. Am J Dis Child 140:543–546

Groothuis JR, Nishida H (2002) Prevention of respiratory syncytial virus infections in high-risk infants by monoclonal antibody (palivizumab). Pediatr Int 44:235–241

Groothuis JR, Levin MJ, Rodriguez W, Hall CB, Long CE, Kim HW, Lauer BA, Hemming VG, The RSVIG Study Group (1991) Use of intravenous gamma globulin to passively immunize high-risk children against RSV: safety and pharmacokinetics. Antimicrob Agents Chemother 35:1469–1473

Groothuis JR, Simoes EA, Levin MJ, Hall CB, Long CE, Rodriguez WJ, Arrobio J, Meissner HC, Fulton DR, Welliver RC, The Respiratory Syncytial Virus Immune Globulin Study Group (1993) Prophylactic administration of respiratory syncytial virus immune globulin to high-risk infants and young children. N Engl J Med 329:1524–1530

Hemming VG, Prince GA, Horswood RL, London WT, Murphy BR, Walsh EE, Fischer GW, Weisman LE, Baron PA, Chanock RM (1985) Studies of passive immunotherapy for infections of respiratory syncytial virus in the respiratory tract of a primate model. J Infect Dis 152:1083–1087

Henderson FW, Collier AM, Clyde WA Jr, Denny FW (1979) Respiratory-syncytial-virus infections, reinfections and immunity. A prospective, longitudinal study in young children. N Engl J Med 300:530–534

Hieter PA, Maizel JV Jr, Leder P (1982) Evolution of human immunoglobulin kappa J region genes. J Biol Chem 257:1516–1522

Johnson PR, Spriggs MK, Olmsted RA, Collins PL (1987) The G glycoprotein of human respiratory syncytial viruses of subgroups A and B: extensive sequence divergence between antigenically related proteins. Proc Natl Acad Sci USA 84:5625–5629

Johnson S, Oliver C, Prince GA, Hemming VG, Pfarr DS, Wang SC, Dormitzer M, O'Grady J, Koenig S, Tamura JK, Woods R, Bansal G, Couchenour D, Tsao E, Hall WC, Young JF (1997) Development of a humanized monoclonal antibody (MEDI-493) with potent in vitro and in vivo activity against respiratory syncytial virus. J Infect Dis 176:1215–1224

Johnson S, Griego SD, Pfarr DS, Doyle ML, Woods R, Carlin D, Prince GA, Koenig S, Young JF, Dillon SB (1999) A direct comparison of the activities of two humanized respiratory syncytial virus monoclonal antibodies: MEDI-493 and RSHZl9. J Infect Dis 180:35–40

Junghans RP (1997) Finally! The Brambell receptor (FcRB). Mediator of transmission of immunity and protection from catabolism for IgG. Immunol Res 16:29–57

Kabat EA, Wu TT, Perry HM, Gottesman KS, Foeller C (1991) Sequences of proteins of immunological interest. U.S. Department of Health and Human Services, Washington, DC

Kapikian AZ, Mitchell RH, Chanock RM, Shvedoff RA, Stewart CE (1969) An epidemiologic study of altered clinical reactivity to respiratory syncytial (RS) virus infection in children previously vaccinated with an inactivated RS virus vaccine. Am J Epidemiol 89:405–421

Kasel JA, Walsh EE, Frank AL, Baxter BD, Taber LH, Glezen WP (1987) Relation of serum antibody to glycoproteins of respiratory syncytial virus with immunity to infection in children. Viral Immunol 1:199–205

Lamprecht CL, Krause HE, Mufson MA (1976) Role of maternal antibody in pneumonia and bronchiolitis due to respiratory syncytial virus. J Infect Dis 134:211–217

Lopez JA, Villanueva N, Melero JA, Portela A (1988) Nucleotide sequence of the fusion and phosphoprotein genes of human respiratory syncytial (RS) virus Long strain: evidence of subtype genetic heterogeneity. Virus Res 10:249–261

Lopez JA, Penas C, Garcia-Barreno B, Melero JA, Portela A (1990) Location of a highly conserved neutralizing epitope in the F glycoprotein of human respiratory syncytial virus. J Virol 64:927–930

Lopez JA, Bustos R, Orvell C, Berois M, Arbiza J, Garcia-Barreno B, Melero JA (1998) Antigenic structure of human respiratory syncytial virus fusion glycoprotein. J Virol 72:6922–6928

Martin-Gallardo A, Fien KA, Hu BT, Farley JF, Seid R, Collins PL, Hildreth SW, Paradiso PR (1991) Expression of the F glycoprotein gene from human respiratory syncytial virus in Escherichia coli: mapping of a fusion inhibiting epitope. Virology 184:428–432

Meissner HC, Fulton DR, Groothuis JR, Geggel RL, Marx GR, Hemming VG, Hougen T, Snydman DR (1993) Controlled trial to evaluate protection of high-risk infants against RSV disease by using standard intravenous immune globulin. Antimicrob Agents Chemother 37:1655–1658

Merz DC, Scheid A, Choppin PW (1980) Importance of antibodies to the fusion glycoprotein of paramyxoviruses in the prevention of spread of infection. J Exp Med 151:275–288

Murphy BR, Walsh EE (1988) Formalin-inactivated respiratory syncytial virus vaccine induces antibodies to the fusion glycoprotein that are deficient in fusion-inhibiting activity. J Clin Microbiol 26:1595–1597

Porterfield JS (1986) Antibody-dependent enhancement of viral infectivity. Adv Virus Res 31:335–355

Prince GA, Hemming VG, Horswood CL, Chanock RM (1985a) Immunoprophylaxis and immunotherapy of respiratory syncytial virus infection in the cotton rat. Virus Res 3:193–206

Prince GA, Horswood RL, Chanock RM (1985b) Quantitative aspects of passive immunity to respiratory syncytial virus infection in infant cotton rats. J Virol 55:517–520

Rimensberger PC, Burek-Kozlowska A, Morell A, Germann D, Eigenmann AK, Steiner F, Burger R, Kuenzli M, Schaad UB (1996) Aerosolized immunoglobulin treatment of respiratory syncytial virus infection in infants. Pediatr Infect Dis J 15:209–216

Rodriguez WJ, Gruber WC, Groothuis JR, Simoes EA, Rosas AJ, Lepow M, Kramer A, Hemming V (1997) Respiratory syncytial virus immune globulin treatment of RSV lower respiratory tract infection in previously healthy children. Pediatrics 100:937–942

Saez-Llorens X, Castano E, Null D, Steichen J, Sanchez PJ, Ramilo O, Top FH Jr, Connor E (1998) Safety and pharmacokinetics of an intramuscular humanized monoclonal antibody to respiratory syncytial virus in premature infants and infants with bronchopulmonary dysplasia. The MEDI-493 Study Group. Pediatr Infect Dis J 17:787–791

Schenerman MA, Hope JN, Kletke C, Singh JK, Kimura R, Tsao EI, Folena-Wasserman G (1999) Comparability testing of a humanized monoclonal antibody (Synagis) to support cell line stability, process validation, and scale-up for manufacturing. Biologicals 27:203–215

Siber GR, Leszcynski J, Pena-Cruz V, Ferren-Gardner C, Anderson R, Hemming VG, Walsh EE, Burns J, McIntosh K, Gonin R, Anderson LJ (1992) Protective activity of a human respiratory syncytial virus immune globulin prepared from donors screened by microneutralization assay. J Infect Dis 165:456–463

Simoes EA, Sondheimer HM, Top FH Jr, Meissner HC, Welliver RC, Kramer AA, Groothuis JR (1998) Respiratory syncytial virus immune globulin for prophylaxis against respiratory syncytial virus disease in infants and children with congenital heart disease. The Cardiac Study Group. J Pediatr 133:492–499

Smith BJ, Lawrence MC, Colman PM (2002) Modelling the structure of the fusion protein from human respiratory syncytial virus. Protein Eng 15:365–371

Subramanian KN, Weisman LE, Rhodes T, Ariagno R, Sanchez PJ, Steichen J, Givner LB, Jennings TL, Top FH Jr, Carlin D, Connor E (1998) Safety, tolerance and pharmacokinetics of a humanized monoclonal antibody to respiratory syncytial virus in premature infants and infants with bronchopulmonary dysplasia. MEDI-493 Study Group. Pediatr Infect Dis J 17:110–115

Taylor GE, Stott EJ, Bew M, Fernie BF, Cote PJ, Collins AP, Hughes M, Jebbert J (1984) Monoclonal antibodies protect against respiratory syncytial virus infection in mice. Immunology 52:137–141

Tempest PR, Bremner P, Lambert M, Taylor G, Furze JM, Carr FJ, Harris WJ (1991) Reshaping a human monoclonal antibody to inhibit human respiratory syncytial virus infection in vivo. Biotechnology (N Y) 9:266–271

The IMpact-RSV Study Group (1998) Palivizumab, a humanized respiratory syncytial virus monoclonal antibody, reduces hospitalization from respiratory syncytial virus infection in high-risk infants. Pediatrics 102:531–537

The PREVENT Study Group (1997) Reduction of respiratory syncytial virus hospitalization among premature infants and infants with bronchopulmonary dysplasia using respiratory syncytial virus immune globulin prophylaxis. Pediatrics 99:93–99

Trudel M, Nadon F, Seguin C, Dionne G, Lacroix M (1987) Identification of a synthetic peptide as part of a major neutralization epitope of respiratory syncytial virus. J Gen Virol 68:2273–2280

Walsh EE, Hruska J (1983) Monoclonal antibodies to respiratory syncytial virus proteins: identification of the fusion protein. J Virol 1983 47:171–177

Walsh EE, Schlesinger JJ, Brandriss MW (1984) Protection from respiratory syncytial virus infection in cotton rats by passive transfer of monoclonal antibodies. Infect Immun 43:756–758

Weltzin R, Traina-Dorge V, Soike K, Zhang JY, Mack P, Soman G, Drabik G, Monath TP (1996) Intranasal monoclonal IgA antibody to respiratory syncytial virus protects rhesus monkeys against upper and lower respiratory tract infection. J Infect Dis 174:256–261

Wright PF, Shinozaki T, Fleet W, Sell SH, Thompson J, Karzon DT (1976) Evaluation of a live, attenuated respiratory syncytial virus vaccine in infants. J Pediatr 88:931–936

Wright PF, Belshe RB, Kim HW, Van Voris LP, Chanock RM (1982) Administration of a highly attenuated, live respiratory syncytial virus vaccine to adults and children. Infect Immun 37:397–400

Wu H, An LL (2003) Tailoring kinetics of antibodies using focused combinatorial libraries. In: Welschof M, Krauss J (eds) Recombinant antibodies for cancer therapy: methods and protocols (Methods in molecular biology, vol. 207). Humana Press, Totowa, pp 213–233

Wu H, Beuerlein G, Nie Y, Smith H, Lee BA, Hensler M, Huse WD, Watkins JD (1998) Stepwise in vitro affinity maturation of Vitaxin, an α v β3-specific humanized mAb. Proc Natl Acad Sci USA 95:6037–6042

Wu H, Pfarr DS, Tang Y, An LL, Patel NK, Watkins JD, Huse WD, Kiener PA, Young JF (2005) Ultra-potent antibodies against respiratory syncytial virus: effects of binding kinetics and binding valence on viral neutralization. J Mol Biol 350:126–144

Wu H, Pfarr DS, Johnson S, Brewah YA, Woods RM, Patel NK, White WI, Young JF, Kiener PA (2007) Development of motavizumab, an ultra-potent antibody for the prevention of respiratory syncytial virus infection in the upper and lower respiratory tract. J Mol Biol 368:652–665

Zhao X, Sullender WM (2005) In vivo selection of respiratory syncytial viruses resistant to palivizumab. J Virol 79:3962–3968

Zhao X, Chen FP, Sullender WM (2004) Respiratory syncytial virus escape mutant derived in vitro resists palivizumab prophylaxis in cotton rats. Virology 318:608–612

Zhao X, Liu E, Chen FP, Sullender WM (2006) In vitro and in vivo fitness of respiratory syncytial virus monoclonal antibody escape mutants. J Virol 80:11651–11657

The Molecular Basis of Antibody Protection Against West Nile Virus

E. Mehlhop and M.S. Diamond(✉)

Abstract West Nile virus (WNV) infection of mosquitoes, birds, and vertebrates continues to spread in the Western Hemisphere. In humans, WNV infects the central nervous system and causes severe disease, primarily in the immunocompromised and elderly. In this review we discuss the mechanisms by which antibody controls WNV infection. Recent virologic, immunologic, and structural experiments have enhanced our understanding on how antibodies neutralize WNV and protect against disease. These advances have significant implications for the development of novel antibody-based therapies and targeted vaccines.

M.S. Diamond

Departments of Medicine, Molecular Microbiology, Pathology & Immunology, Washington University School of Medicine, 660 South Euclid Avenue, Box 8051, St. Louis, MO 63110, USA
e-mail: diamond@borcim.wustl.edu

S.K. Dessain (ed.) *Human Antibody Therapeutics for Viral Disease. Current Topics in Microbiology and Immunology 317.*
© Springer-Verlag Berlin Heidelberg 2008

1   West Nile Virus

West Nile virus (WNV) is an enveloped, arthropod-borne RNA virus of the Flaviviridae family. Isolated in 1937 in the West Nile district of Uganda from a woman with an undiagnosed febrile illness (Smithburn et al. 1940), WNV is closely related to a group of viruses that cause disease globally. The genus *Flavivirus* is composed of 73 viruses, approx. 40 of which are associated with severe human diseases, including dengue (DENV), yellow fever (YFV), Japanese encephalitis (JEV), and tick-borne encephalitis (TBE) viruses (Burke et al. 2001). Based on antibody cross-neutralization patterns, WNV shares a serocomplex with the closely related JEV, Saint Louis (SLEV), and Murray Valley (MVEV) encephalitis viruses (Brinton 2002; Lindenbach et al. 2001). WNV cycles enzootically between *Culex* mosquitoes and birds, although it also can infect and cause illness in a range of vertebrate species including humans and horses (Hayes et al. 2005a; Kile et al. 2005; Miller et al. 2003). These vertebrate animals act as dead-end hosts and generally do not contribute to virus spread or evolution in nature because infection in nonavian species results in a low-level, transient viremia that is generally insufficient for infection of mosquitoes (Hubalek et al. 1999a).

Historically, WNV caused sporadic outbreaks of a mild febrile illness in regions of Africa, the Middle East, Asia, and Australia. However, in the 1990s the epidemiology of infection appeared to change. New outbreaks in parts of Eastern Europe were associated with higher rates of severe neurological disease (Hubalek et al. 1999a, b). In 1999, WNV entered North America and caused seven human fatalities in the New York area as well as the deaths of a large number of birds and horses. Over the last 7 years WNV has spread throughout the continental United States, as well as to parts of Canada, Mexico, the Caribbean, and Central and South America (Deardorff et al. 2006; Komar et al. 2006). Nucleotide sequencing separates WNV strains into two major lineages (Lanciotti et al. 2002). Lineage I viruses are emerging globally, and subsets of these strains have been associated with severe human and avian disease (Beasley et al. 2002b; Brault et al. 2004). In contrast, lineage II viruses isolated from central and southern Africa and parts of Asia have not been associated with severe human disease (Berthet et al. 1997; Jupp 2001). Because of the increased range of WNV, the number of human cases has continued to rise. During an epidemic the seroconversion rate within the affected human population is estimated at approx. 3% (Mostashari et al. 2001; Petersen et al. 2003; Tsai et al. 1998) and the incidence of severe disease is approx. 7/100,000 (Huhn et al. 2005). Overall, only a small percentage of humans (1/150) with WNV infection develop severe neurological disease, which can include cognitive dysfunction, ocular manifestations, meningitis, encephalitis, and flaccid paralysis (Hayes et al. 2005b; Petersen et al. 2003; Sejvar et al. 2003). Neuroinvasive WNV infection results in roughly 10% mortality, and long-term neurological dysfunction is commonly observed (Davis et al. 2006c; Sejvar et al. 2006). In the United States between 1999 and 2006, approx. 23,000 cases were diagnosed and associated with 880 deaths. However, the spectrum of disease may be much larger. In 2003 alone, based on screening of blood-bank samples, there were an estimated 730,000 undiagnosed infections (Busch et al. 2005). No vaccines or specific therapies for WNV infection are currently approved for human use.

WNV is composed of an approx. 11-kb single-stranded, positive-polarity RNA genome encoding a single open reading frame (ORF) bracketed by 5′ and 3′ nontranslated regions (NTRs). The NTRs contain conserved sequences and predicted secondary structures that provide signals for negative strand synthesis, genome amplification, translation, and packaging (Lindenbach et al. 2001). WNV genomic RNA is translated as a single polyprotein that is both co- and post-translationally cleaved by host and viral proteases into three structural and seven nonstructural (NS) proteins (Nowak et al. 1989). The N-terminal one-third of the WNV genome encodes the viral structural proteins capsid (C), membrane (prM/M), and envelope (E), which are involved in viral assembly, host cell binding, and entry. The remaining C-terminal region encodes the seven NS proteins (NS1, NS2A/B, NS3, NS4A/B, and NS5) involved in viral transcription, RNA replication, and attenuation of host antiviral responses. NS1 is a cofactor activity for the viral replicase (Khromykh et al. 2000; Lindenbach et al. 1999), it is secreted from infected cells (Winkler et al. 1988; Winkler et al. 1989), and recent studies suggest it may have immune-evasive function (Chung et al. 2006a). NS2A inhibits interferon (IFN) responses and may participate in virus assembly (Liu et al. 2004, 2005, 2006; Munoz-Jordan et al. 2003). NS2B is a cofactor required for NS3 proteolytic activity and contributes to antagonism of IFN responses (Liu et al. 2005). NS3 has protease, NTPase (nucleoside 5′ triphosphatase), and helicase activities (Khromykh et al. 2000; Liu et al. 2003). NS4A and NS4B can block IFN-mediated signal transducer and activator of transcription (STAT) signaling (Liu et al. 2006; Munoz-Jordan et al. 2003, 2005). NS5 encodes the RNA-dependent RNA polymerase and methyltransferase, and may independently antagonize antiviral IFN responses (Best et al. 2005; Egloff et al. 2002; Khromykh et al. 1999; Lin et al. 2006; Yusof et al. 2000).

WNV infection occurs following cellular attachment and receptor-mediated endocytosis. Although both dendritic cell-specific intercellular adhesion molecule 3 grabbing nonintegrin (DC-SIGN)-R and the $\alpha_v\beta_3$ integrin have been implicated as WNV attachment ligands (Chu et al. 2004a; Davis et al. 2006b), the cellular receptors for WNV on physiologically relevant cell types such as neurons remains poorly understood. Cellular entry appears to require the formation of clathrin-coated pits (Chu et al. 2004b). Following a pH-dependent conformational change in the E protein (Modis et al. 2004; Zhang et al. 2004), the viral and endosomal membranes fuse, releasing the viral nucleocapsid into the cytoplasm (Allison et al. 1995; Gollins et al. 1986a). Upon nucleocapsid release, viral RNA associates with endoplasmic reticulum (ER) membranes and is translated. Translation is a prerequisite for generating a negative-strand RNA intermediate that serves as a template for nascent positive-strand genomic RNA synthesis (Mackenzie et al. 2001). Flaviviral RNA synthesis is semi-conservative and asymmetric, as positive-strand RNA genome production is about ten times more efficient than negative-strand synthesis (Brinton 2002). Positive-strand RNA is either packaged within progeny virions or used to translate additional viral proteins. WNV assembles and buds into the ER to form enveloped immature particles containing the prM protein. Following transport through the trans-Golgi network, furin-mediated cleavage of prM to M generates mature, 500-Å infectious virions that are released by exocytosis (Elshuber et al. 2003; Guirakhoo et al. 1992; Stadler et al. 1997).

WNV is an emerging human and animal pathogen and a member of a family of viruses that causes significant disease. Intensive study has increased our understanding of WNV pathogenesis and the immune responses required for protection from severe neuroinvasive disease (Lim et al. 2006; Samuel et al. 2006a; Wang et al. 2004a). This review will focus on the recent advances in understanding the role of antibody in protection from disease, the mechanisms of antibody-mediated WNV neutralization, and the potential for development of vaccines and antibodies for therapeutic use.

2   Pathogenesis and Immune Protection

The pathogenesis of WNV and the immune responses that prevent central nervous system (CNS) dissemination are beginning to be characterized through studies in small animal models. Following peripheral inoculation WNV is believed to initially replicate in skin dendritic cells (DCs) (Johnston et al. 1996, 2000). These cells then migrate to the draining lymph nodes (Byrne et al. 2001; Johnston et al. 2000), where a new round of infection occurs that leads to a subsequent viremia and spread to the visceral organs (e.g., kidney and spleen). Dissemination of WNV to the CNS occurs shortly before clearance of infectious virus from peripheral tissues (Diamond et al. 2003a). Infectious WNV is detected within the CNS at multiple sites including the cerebral cortex, hippocampus, basal ganglia, cerebellum, brain stem, and spinal cord (Chambers et al. 2003; Diamond et al. 2003a; Xiao et al. 2001). In most animals, CNS infection occurs primarily in neurons and is associated with their degeneration, loss of cell architecture, and apoptosis (Shrestha et al. 2003). However, the mechanisms of WNV CNS seeding remain poorly understood. Earlier entry of WNV in the CNS has been observed in mice exhibiting increased levels of viremia, suggesting hematogenous spread may contribute to CNS seeding (Diamond et al. 2003c; Samuel et al. 2005). Yet, earlier viral invasion of the CNS has also been seen in complement-deficient mice that have normal levels of viremia (Mehlhop et al. 2005, 2006). These studies suggest that additional soluble inflammatory factors may affect blood–brain barrier (BBB) permeability and WNV CNS seeding. Indeed, recent evidence has suggested tumor necrosis factor (TNF)-α-mediated changes in BBB permeability can enhance WNV CNS seeding (Wang et al. 2004b). Retrograde neuronal transport may also contribute to CNS dissemination (Hunsperger et al. 2006). Clearly, additional studies are necessary to define the precise mechanism(s) for spread of WNV to the CNS.

Experiments in small animals suggest that both innate and adaptive immune responses are required to control WNV dissemination and disease. Type I IFN (α/β) and its downstream effector molecules protein kinase (PK)R and RNAse L are critical components of the innate immune response to WNV infection (Samuel et al. 2005, 2006b; Scherbik et al. 2006). Pretreatment of cells with type I IFN in vitro prevents WNV infection (Anderson et al. 2002; Fredericksen et al. 2004; Keller et al. 2006; Lucas et al. 2003; Samuel et al. 2005). Analogously, a deficiency of type I IFN signaling in vivo results in increased viral replication, expanded tropism, and uniform lethality (Keller et al. 2006; Liu et al. 2005; Samuel et al. 2005). Similarly, type II IFN (IFN-γ) produced by $\gamma\delta$ T cells also limits peripheral viral replication and early WNV dissemination into the CNS (Shrestha et al. 2006b; Wang et al. 2003a). Additional innate immune responses, including toll-like receptor 3 (TLR3) and $2'5'$ oligoadenylate synthetase activities, also regulate WNV infection in vivo (Mashimo et al. 2002; Perelygin et al. 2002; Wang et al. 2004b). Development of antiviral adaptive immunity is also necessary for protection from disease, as passive transfer of immune antibody protected wild-type and B cell-deficient mice from lethal WNV challenge. However, antibody alone did not eradicate infection in *RAG1*-deficient hosts, which lack both B and T cells (Diamond et al. 2003a; Engle et al. 2003), indicating T cells likely have a critical role limiting severe WNV disease. Indeed, mice lacking either B cells or CD8+ T cells exhibited increased viral burden and lethality following peripheral WNV infection (Diamond et al. 2003a; Shrestha et al. 2004; Wang et al. 2003b). Cytolytic T cell responses are required for clearance of WNV infection, as persistence within the CNS was observed in mice that lacked either classical class I MHC or perforin molecules (Shrestha et al. 2006a; Wang et al. 2004c). Antibody responses may also contribute to viral clearance in the CNS (Griffin et al. 2001; Levine et al. 1991), as passive transfer of a potently neutralizing mAb against WNV after CNS infection protected mice and hamsters against lethal infection (Morrey et al. 2006; Oliphant et al. 2005).

The priming of early effective neutralizing antiviral antibody responses is crucial for control of severe WNV infection. In wild-type C57BL/6 mice, the development of WNV-specific neutralizing IgM was consistently observed beginning on day 4 after subcutaneous infection (Fig. 1; Diamond et al. 2003a; Mehlhop et al. 2005, 2006). Mice lacking secreted IgM (sIgM$^{-/-}$) were highly susceptible to lethal WNV infection and exhibited sustained viremia, earlier viral entry into the CNS, and greater CNS viral accumulation (Diamond et al. 2003c). Transfer of serum from wild-type to sIgM$^{-/-}$ mice on day 4 after infection significantly protected mice from lethal WNV infection. This suggested that early IgM neutralizing antibodies were critical for control of WNV disease. Indeed, the level of WNV-specific IgM in serum on day 4 after infection predicted disease outcome in mice (Diamond et al. 2003c). Accordingly, immune deficiencies that impair antibody priming also predispose to WNV susceptibility. Mice lacking the C3 or C4 components of complement, or complement receptors 1 and 2, exhibited blunted antiviral antibody priming and enhanced susceptibility to lethal WNV infection (Mehlhop et al. 2005, 2006). Additionally, the absence of CD4+ T cells, class II MHC expression, or CD40 signaling decreased neutralizing antiviral antibody responses and survival rates after WNV infection (Sitati and Diamond 2006; Sitati et al. 2007).

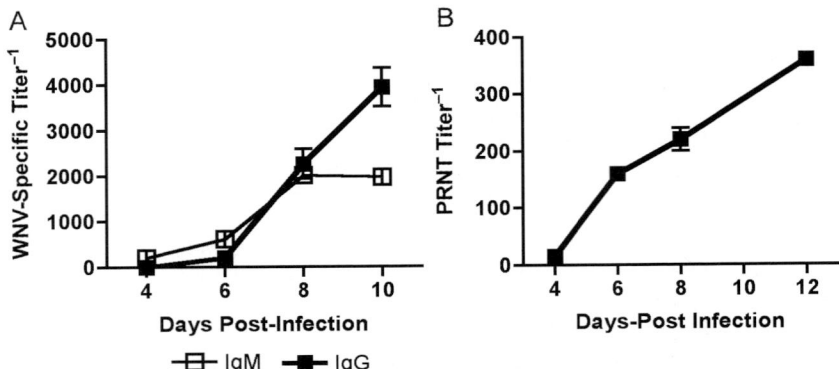

Fig. 1 WNV antibody responses in C57BL/6 mice. The kinetics WNV-specific neutralizing antibody development are shown following subcutaneous inoculation of 8- to 10-week-old C57BL/6 mice with 10^2 plaque-forming units (PFU) of WNV. **A** Endpoint titers of WNV-specific IgM and IgG were determined by ELISA. **B** WNV neutralizing activity of heat-inactivated serum was assayed by classical plaque reduction neutralization (PRNT) assay

3 Structural Determinants of Antibody-Mediated WNV Neutralization

The E glycoprotein is the major surface protein on the *Flavivirus* virion and is the principal antigen that elicits protective neutralizing antibodies (Roehrig et al. 2001). However, a subset of neutralizing antibodies to flaviviruses may also recognize the prM protein on the virion (Churdboonchart et al. 1991; Vazquez et al. 2002). Interestingly, antibodies to the nonstructural protein NS1, which is absent from the virion, also are protective in vivo (Chung et al. 2006b). Antibody responses to NS3 and NS5 have also been observed during WNV infection (Churdboonchart et al. 1991; Valdes et al. 2000; Wong et al. 2003), although their functional significance remains uncertain.

3.1 E Protein Structure and WNV Virion Composition

Recent studies of *Flavivirus* E protein structure have elucidated critical determinants of antibody-mediated protection. The crystal structure of soluble E protein has been determined for DENV, tick-borne encephalitis virus (TBEV), and WNV (Bressanelli et al. 2004; Kanai et al. 2006; Modis et al. 2003, 2004, 2005; Nybakken et al. 2006; Rey et al. 1995, 2003; Zhang et al. 2004). The E protein has three structural domains that mediate viral attachment, entry, and viral assembly (Fig. 2A). Domain I (DI), the central structural domain, is an 8-stranded β-barrel that contributes to the conformational changes that occur after exposure to acid pH in the endosome.

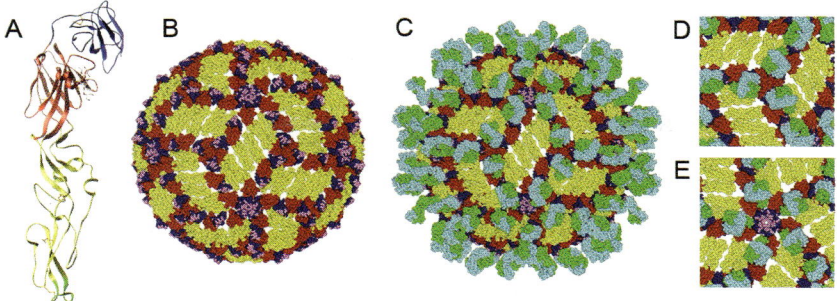

Fig. 2 WNV E protein structure and neutralizing Fab fragment binding of the virion. **A** DI (*red*), DII (*yellow*), and DIII (*blue*) domains of monomeric WNV E protein as determined by X-ray crystallography. The fusion loop (residues 98–110) is highlighted in *green*. **B** Pseudoatomic model of the cryoelectron microscopic reconstruction of the WNV virion. The E16 structural epitope is mapped in *magenta*. **C** Saturation binding of E16 on the WNV particle. E16 is predicted to bind 120 out of 180 potential epitopes with exclusion from the inner fivefold axis. **D** and **E** Magnified regions of the boxed areas in panel **C**. (Reprinted with permission from Macmillan Publishers Ltd. For details, see Nybakken et al. 2005, copyright 2005)

Domain II (DII) is a 12-stranded β-barrel that is involved in dimerization (Rey et al. 1995) and contains a highly conserved hydrophobic fusion-peptide that mediates the class II acid-catalyzed fusion event (Allison et al. 2001; Bressanelli et al. 2004; Modis et al. 2004). Domain III (DIII) adopts an immunoglobulin-like fold and contains a putative receptor-binding domain (Bhardwaj et al. 2001; Chu et al. 2005; Rey et al. 1995). Short, flexible linkers connect the E protein domains and allow for the conformational changes associated with virus maturation and membrane fusion. Several unique characteristics of WNV E protein relative to other flaviviruses have been identified by X-ray crystallography (Kanai et al. 2006; Nybakken et al. 2006). In contrast to the dimeric forms of DENV and TBEV E proteins, WNV E protein exists as a monomer in solution. Differences in the angle observed at the DI–DII hinge between WNV and DENV suggest critical WNV E dimer contact residues may be weaker or more easily disrupted. Nonetheless, WNV E proteins form dimers on the surface of the mature virion (Mukhopadhyay et al. 2003). Additionally, the single N-linked glycosylation site on E, which is essential for DC-SIGN-R recognition (Davis et al. 2006b) and neurovirulence of WNV (Beasley et al. 2005; Shirato et al. 2004), is located on a unique α-helical segment in DI of WNV E in comparison to other flaviviruses. Differences in the location of the glycosylation site may contribute to differences in viral tropism and pathogenesis among flaviviruses (Davis et al. 2006a; Hanna et al. 2005).

Understanding E protein structure in the context of WNV virion assembly has provided fundamental insights into the potential mechanisms of antibody-mediated neutralization. The glycoproteins on the surface of the 600-Å immature *Flavivirus* virion are organized into 60 asymmetric trimeric spikes of prM-E heterodimers (Zhang et al. 2003a, b). At the apices of the spikes, prM caps the fusion loop of E

(Roehrig et al. 1990), presumably to prevent premature fusion as the virus passes through the acidic secretory pathway (Heinz et al. 1994a, b). A membrane proximal furin-catalyzed cleavage releases the N-terminal pre-peptide from prM (Stadler et al. 1997; Wengler 1989) allowing the transition from trimeric prM-E heterodimers to E homodimers found in the mature 500-Å enveloped virion (Mukhopadhyay et al. 2003; Zhang et al. 2003b, 2004). Cryoelectron microscopy (cryoEM) has shown that in the mature virus head-to-tail homodimers of E form a smooth icosahedral protein shell over the lipid bilayer in a "herringbone" pattern that defines three repeating environments. The twofold, threefold, and fivefold axes of symmetry are defined by the dimerization of E, and radial arrangement of DI and DIII, respectively (Kuhn et al. 2002; Mukhopadhyay et al. 2003). Antibody recognition or receptor binding may occur in different symmetry environments, resulting in differential occupancy of the virion. This likely has functional consequences for the recognition of viral particles by different cell types and the immune system.

3.2   Identification of Neutralizing Epitopes

The antigenic domains of *Flavivirus* E proteins were initially characterized by mapping and competition experiments with monoclonal antibodies (mAbs) (reviewed in Roehrig et al. 2001). These studies identified three antigenic domains (C, A, and B), which were later correlated with the structural domains DI, DII, and DIII of *Flavivirus* E proteins (Heinz et al. 1983; Kimura-Kuroda et al. 1983; Roehrig et al. 1998). Many of the B domain epitopes of DENV-2, JEV, and TBE elicited neutralizing virus-specific antibody responses (Heinz et al. 1983; Lin et al. 1994; Mason et al. 1989; Roehrig et al. 1998). However, not all E glycoprotein-reactive antibodies neutralize virus infectivity. Indeed, some virus-specific nonneutralizing mAbs were found to recognize C domain epitopes of DENV recombinant proteins (Megret et al. 1992). Additionally, mAb recognition of TBE A domain epitopes were inhibited by low pH treatment, whereas recognition of B domain epitopes were unaffected (Guirakhoo et al. 1989; Heinz et al. 1983). These studies suggest distinct domains of WNV E protein elicit antibodies with distinct functional activities.

Sequencing of neutralization escape mutants identified DIII as a major target of mAb-mediated flavivirus neutralization (Beasley et al. 2001; Beasley et al. 2002a; Cecilia et al. 1991; Choi et al. 2007; Lin et al. 1994; Roehrig et al. 1998). More recent studies have confirmed the epitopes on DIII responsible for eliciting potently neutralizing antibodies (Table 1). Using both forward and reverse genetic strategies, several groups have established that the most potent WNV neutralizing mAbs bind to the distal lateral ridge of DIII, with key contacts to residues K307, T330, and T332 (Beasley et al. 2002a; Choi et al. 2007; Oliphant et al. 2005; Sanchez et al. 2005; Volk et al. 2004). Neutralizing mAbs that recognize the same residues were also characterized by NMR (Volk et al. 2004; Wu et al. 2003) and X-ray crystallography (Fig. 2B; Nybakken et al. 2005). The latter structural studies

Table 1 Epitope mapping of WNV E protein domain III neutralizing mAbs

mAb	Isotype	E contact residues	Method	Reference
A		307	Neutralization escape	Chambers et al. 1998
5C5	IgG2a	307, 332	Neutralization escape	Beasley et al. 2002
			rDIII expression	Volk el al. 2004
				Choi et al. 2007
7H2	IgG2a	307, 332	Neutralization escape	Beasley et al. 2002
			rDIII expression	Volk el al. 2004
				Choi et al. 2007
3A3	IgG2a	307, 330, 332	Neutralization escape	Beasley et al. 2002
			rDIII expression	Volk el al. 2004
				Choi et al. 2007
5H10	IgG2a	307, 330, 332	Neutralization escape	Beasley et al. 2002
			rDIII expression	Volk el al. 2004
				Choi et al. 2007
E16	IgG2b	302, 306, 307, 308, 309, 330, 331, 332, 333, 365, 366, 367, 368, 389, 390, 391	Yeast display	Oliphant et al. 2005
			X-ray crystallography	Nybakken et al. 2005
E24	IgG2a	307, 330, 332	Yeast display	Oliphant et al. 2005
E27	IgG1	306, 307	Yeast display	Oliphant et al. 2005
E33	IgG2b	307, 330, 332	Yeast display	Oliphant et al. 2005
E40	IgG2a	306, 307	Yeast display	Oliphant et al. 2005
E34	IgG1	307, 330	Yeast display	Oliphant et al. 2005
E43	IgG2a	307, 330	Yeast display	Oliphant et al. 2005
E47	IgG2a	307, 330	Yeast display	Oliphant et al. 2005
E48	IgG2a	306, 307, 330, 332, 381	Yeast display	Oliphant et al. 2005
E58	IgG2a	307, 330, 332	Yeast display	Oliphant et al. 2005
7G9	IgG3	280–466	rE expression	Razumov et al. 2005
11G3	IgG3	280–466	rE expression	Razumov et al. 2005
7E6	IgG3	280–466	rE expression	Razumov et al. 2005
9E2	IgG2a	280–466	rE expression	Razumov et al. 2005
8B10	IgG1	307, 330	VLP mutagenesis	Sanchez et al. 2005
10C5	IgG1	307, 330	VLP mutagenesis	Sanchez et al. 2005
17C8	IgG1	306, 307, 330	VLP mutagenesis	Sanchez et al. 2005
11C2	IgG1	307, 330, 332	VLP mutagenesis	Sanchez et al. 2005
5E8	IgG1	307, 332, 367	Neutralization escape rDIII expression	Choi et al. 2007
CR4299	hIgG1	DIII 7H2 epitope	mAb competition	Throsby et al. 2006
CR4374	hIgG1	DIII 7H2 epitope	mAb competition	Throsby et al. 2006

[a]Indicates the method used to map the mAb contact residues listed

demonstrated that one DIII-specific neutralizing antibody engaged 16 residues in 4 discontinuous regions that localize to the amino terminus (residues 302–309) and 3 strand-connecting loops (residues 330–333, 365–368, and 389–391). Antibody binding at this epitope correlated with potent in vitro neutralization and strong in vivo protection (Oliphant et al. 2005), suggesting this site in DIII may be an important neutralizing epitope. Both Fab fragments and single chain Fv that recognize the DIII lateral ridge epitope neutralize infection, indicating that bivalent cross-linking is not required for DIII-directed antibody-mediated inhibition of WNV infection (Gould et al. 2005). Although DIII has been suggested to contribute to virus attachment, at least some DIII-directed neutralizing mAbs appear to block at a post-attachment step. Potent neutralization was still observed following pre-incubation of cells with WNV prior to the addition of DIII-directed mAb (Nybakken et al. 2005). In contrast, the activity of a neutralizing mAb directed at the fusion peptide in DII was completely lost if virus was bound to cells prior to mAb addition.

Cross-reactive, neutralizing mAbs against flaviviruses generally map to the fusion peptide (amino acids 98–110) in DII (Crill et al. 2004; Oliphant et al. 2006; Stiasny et al. 2006). In one study, 45% (40 of 89) of the DI–DII-specific mAbs showed markedly reduced binding to WNV E protein mutated at the W101 residue in the fusion peptide, and 85% of these (34 of 40) cross-reacted with the distantly related DENV (Oliphant et al. 2006). Other groups also have established that mutations at either G106 or L107 in the fusion peptide eliminate mAb recognition of *Flavivirus* group-specific epitopes (Crill et al. 2004; Stiasny et al. 2006). Additionally, approx. 30% of the cross-reactive antibodies in DENV patient sera mapped to a single amino acid (L107) in the fusion loop (Stiasny et al. 2006). Preliminary studies with human mAbs suggest that the cross-reactive fusion-peptide epitope may be immunodominant, whereas the DIII-specific neutralizing epitope appears less dominant (Gould et al. 2005; Throsby et al. 2006). It is intriguing to consider that flaviviruses may manipulate the humoral response to direct antibody specificity away from highly protective DIII neutralizing epitopes. Definitive studies examining the human antibody repertoire in larger numbers of individuals following WNV infection are needed to resolve this issue.

Less strongly neutralizing WNV-specific mAbs mapped to six additional sites in DI and DII outside of the fusion loop: the lateral ridge of DI, the linker region between DI and DIII, the hinge interface between DI and DII, the lateral ridge, the central interface, and the dimer interface of DII (Oliphant et al. 2006). These mAbs exhibited little neutralization activity by classical plaque reduction assays, but inhibited infection on cells expressing alternate WNV attachment receptors, such as DC-SIGN-R. Interestingly, most DI–DII-specific mAbs still protected mice from lethal WNV challenge (Oliphant et al. 2006), although they were less effective than DIII-specific neutralizing mAbs. These data suggest DIII- and DI–DII-specific mAbs neutralize and protect against WNV infection through independent mechanisms.

Most flavivirus-reactive mAbs have the ability to enhance infection of Fc-γ receptor (FcγR)-bearing cells (Halstead et al. 1977, 1984; Peiris et al. 1979, 1982). This phenomenon, antibody-dependent enhancement (ADE) of infection, has not been documented to contribute to WNV disease, although it has been implicated in the pathogenesis of severe DENV infection (reviewed in Halstead 1988; Morens 1994). WNV DI–DII-directed mAbs appear to behave distinctly from DIII-specific mAbs during infection of FcγR+ cells. While DIII-directed mAb neutralized infection at high concentrations on FcγR+ cells, DI or DII-specific mAbs enhanced infection at similar concentrations (Nybakken et al. 2005; Oliphant et al. 2006). This could reflect an inherent inability of these mAbs to completely neutralize infection. Alternatively, differences in the mechanism of neutralization (attachment versus post-attachment) could explain the distinct enhancement patterns. Indeed, WNV DIII-specific neutralizing mAbs have been suggested to block infection at a post-attachment step (Nybakken et al. 2005), which should inhibit infection regardless of the mechanism of cellular entry. In contrast, binding assays suggest DII fusion-loop specific mAbs may inhibit WNV receptor attachment. Since mAbs can facilitate WNV attachment to $FcγR^+$ cells via the constant region Fc moiety, mAbs that inhibit attachment likely exhibit less neutralizing activity on these cells. Thus, differences in epitope recognition may have different functional consequences depending on the cell type infected.

3.3   Viral Particle Occupancy

In addition to correlating epitope location with mAb function, recent studies have shown differences in occupancy of specific mAbs on WNV virions. Consistent with proposed docking models (Fig. 2C; Nybakken et al. 2005), cryoEM studies found a DIII-directed neutralizing Fab bound a maximum of 120 of 180 available sites on the mature WNV virion (Kauffman et al. 2006). Steric hindrance excluded Fab binding at sites at the inner fivefold symmetry axis, suggesting that binding site saturation is not required for mAb-mediated WNV neutralization. In contrast, no differences in epitope accessibility were predicted for several DII-specific neutralizing mAbs in any of the E protein symmetry environments (Oliphant et al. 2006). However, many DI–DII-directed neutralizing mAbs mapped to sites that were believed to be poorly accessible on the mature WNV virion. Destabilized WNV E dimer contacts could affect exposure of these sites and mAb recognition and neutralization (Oliphant et al. 2006; Stiasny et al. 2006). Interestingly, several of these DI–DII-directed mAbs still exhibited significant neutralizing activity, suggesting partially mature WNV virions that contain accessible DI and DII sites in the context of prM-E heterodimers may exist. Indeed, a recent study indicated that WNV infectious particles may still contain uncleaved prM (Davis et al. 2006b), the presence of which can markedly affect mAb neutralization (Guirakhoo et al. 1992; Heinz et al. 1994b). Alternatively, viral particles may not be as static as the cryoEM pseudo-atomic model suggests.

4   Multiple Neutralizing Mechanisms of WNV Antibodies

Virus neutralization has been extensively studied and distinct mechanisms of mAb-mediated virus neutralization have been proposed (reviewed in Burton et al. 2001; Dimmock 1993; Parren et al. 2001). Antiviral antibody recognition may prevent viral infection by aggregating virions, inhibiting virion attachment, preventing virus-membrane fusion, or hindering viral uncoating. Additionally, mAb binding to virions or virus-infected cells may activate other Fc-dependent antiviral activities, such as complement activation or FcγR-mediated immune complex clearance (Zinkernagel et al. 2001). Recent evidence suggests different neutralizing mAbs may block WNV infection by inhibiting multiple stages of the viral life cycle.

4.1   Virus Neutralization by mAbs

Antibody recognition can prevent *Flavivirus* infection by blocking cellular attachment. Some neutralizing mAbs against DENV inhibit infection only when added before virus adsorption to cells (Crill et al. 2001; Hung et al. 1999). The most potent of these inhibitory mAbs mapped to DIII of DENV E protein (Crill et al. 2001). Similarly, WNV DIII-directed mAbs partially prevented virus attachment to cells (Nybakken et al. 2005). Interestingly, DII-specific mAbs directed at the fusion peptide more strongly inhibited WNV attachment and exhibited little neutralizing activity when added after viral attachment (Nybakken et al. 2005). The apparent disparity between the function of WNV and DENV DIII-specific mAbs could be due to subtle differences in epitope recognition, allowing mAb binding at the five-fold symmetry axis. Indeed, recent cryoEM studies with a DIII-specific DENV2 neutralizing mAb suggest complete (180 of 180) occupancy of E proteins (R. Kuhn, personal communication). By comparing the attachment-blocking activity of DIII-specific mAbs against DENV and WNV with their occupancy in the different symmetry groups on the viral particle, we speculate that fivefold clustered DIII may recognize a specific cellular attachment ligand.

Antibodies can also neutralize by inhibiting membrane fusion and viral uncoating. Some mAbs neutralize JEV or DENV infectivity at a post-attachment step (Butrapet et al. 1998; Se-Thoe et al. 2000), and others efficiently block the low pH-dependent syncytium formation that occurs in *Flavivirus*-infected C6/36 insect cells (Guirakhoo et al. 1992; Roehrig et al. 1998). In seminal studies by Gollins and Porterfield, increasing concentrations of polyclonal immune antibody inhibited the pH-dependent uncoating of WNV as measured by detection of cytoplasmic ^{3}H-uridine labeled RNA (Gollins et al. 1986b). Thus, neutralizing antibodies could inhibit WNV infection by blocking the acid-catalyzed fusion event that occurs at the endosomal membrane. More recently, at least some DIII-specific neutralizing mAbs against WNV have been shown to act in this manner. Infectivity studies suggest these mAbs neutralize WNV infectivity primarily at a post-attachment step (Nybakken et al. 2005) and appear to directly block low pH-dependent plasma membrane fusion (B. Thompson, M. Diamond, and D. Fremont, unpublished data). Additionally, structural studies of DIII-E16 Fab complexes suggest this antibody may block the pH-dependent rearrangement of the E protein by tethering a highly conserved Y302 residue in the DI–DIII linker that makes a required hydrogen bond in the post-fusion structure (Nybakken et al. 2005). Taken together, these data indicate highly potent DIII-specific neutralizing mAbs against WNV likely prevent pH-dependent fusion in the endosome and nucleocapsid escape, thus targeting virions for destruction in the lysosome.

4.2 MAb-Mediated Complement Activation

Antibodies may also inhibit WNV infection by activating Fc-dependent effector functions, including complement activation and Fc-γR targeting. Indeed, virus opsonization with classical pathway complement components C1q, C4b, and C3b inhibit receptor attachment (Berry et al. 1968) and promote the formation of C5b–C9 membrane attack components that induce virolysis (Cooper et al. 1976). Recent studies indicate that complement both directly neutralizes and augments antibody-mediated neutralization of WNV in vitro (Mehlhop et al. 2005). Moreover, in the presence of WNV-specific mAbs, complement promotes lysis of BHK or mouse MC57GL cells that express surface E or NS1 proteins (Mehlhop et al. 2005; K. Chung and M. Diamond, unpublished results). These data suggest antibodies that avidly fix complement may augment neutralizing activity in vivo against WNV and other flaviviruses. Consistent with this, a mouse IgG2a switch variant of an E protein-specific mAb against YFV protected mice more efficiently from lethal infection than the parental IgG1 (Schlesinger et al. 1995). Although complement can augment mAb-mediated neutralization, it also has been implicated in enhancing WNV infection, at least in vitro. Serum complement enhanced CD11b/CD18-dependent infection of WNV by macrophages (Cardosa et al. 1983, 1986). The significance of complement-mediated enhancement of WNV infection in vivo remains unclear.

4.3 Clearance Through Fc-γ Receptor Recognition

Interaction of the antibody Fc region with Fc-γRs contributes to protection against WNV infection in vivo. Mice lacking activating Fc-γRs required significantly higher doses of a neutralizing anti-E mAb to maintain equivalent levels of protection against lethal WNV infection (Oliphant et al. 2005). Similarly, activating Fc-γRs were required for the protective efficacy of nonneutralizing mAbs against the NS1 protein. Passive administration of nonneutralizing mAbs against YFV and DENV NS1 proteins prevented lethal infection (Brandriss et al. 1986; Henchal et al. 1987, 1988; Schlesinger et al. 1985, 1993); this activity required Fc-dependent effector functions, as Fab′$_2$ fragments of anti-YFV NS1 mAbs did not protect in vivo (Schlesinger et al. 1993). The protective effect was independent of complement activation as the efficacy of YFV NS1-specific mAbs were unaffected by depletion of C3 cobra venom factor (Schlesinger et al. 1993). Consistent with this, most NS1-specific mAbs lost protective efficacy against WNV in mice that lacked Fc-γR (Chung et al. 2006b). Although more studies are needed to define the precise protective mechanism, based on depletion studies, the Fc-γR-dependent mAb-mediated effector functions appeared independent of natural killer cells (Chung et al. 2007).

Although interactions with Fc-γR in vivo appear to enhance antibody potency and confer protection against WNV, Fc-γR engagement in vitro can enhance replication of WNV and other flaviviruses in myeloid cells (Gollins et al. 1984, 1985; Halstead 1994; Kliks 1989, 1990; Peiris et al. 1979, 1981). At subneutralizing concentrations or an occupancy that is below the threshold necessary for neutralization, all mAbs that recognize WNV virions will enhance infection in Fc-γR+ cells (T. Pierson and M. Diamond, unpublished results). However, despite its extensive characterization in vitro and its possible epidemiological link to the pathogenesis of Dengue hemorrhagic fever, the pathogenic significance of ADE for WNV and other flaviviruses in vivo remains uncertain (Barrett et al. 1986; Gould et al. 1987, 1989; Wallace et al. 2003). It is also possible that Fc-γR-dependent enhancement of infection of myeloid cells could have beneficial effects in vivo by promoting antigen presentation. FcγR engagement by antibody-opsonized antigen can stimulate dendritic cell and macrophage maturation (Edwards et al. 2006; Nimmerjahn et al. 2005, 2006; Regnault et al. 1999) and promote antigen presentation to both CD4+ and CD8+ T cells (den Haan et al. 2002; Hamano et al. 2000; Regnault et al. 1999).

5   WNV Vaccines and Protective Antibody Responses

Induction of neutralizing antibodies is a critical determinant for the efficacy of WNV vaccines. Several different immunization strategies have been utilized to promote neutralizing antibody responses against WNV in multiple animal models. The development of inhibitory antibodies correlates well with long-term protection of animals from WNV challenge. One major challenge to WNV vaccination may be to induce protective humoral immunity in the populations most at risk for severe clinical disease, the elderly and immunosuppressed (Diamond et al. 2003b; Hayes et al. 2005b). For example, as studies with influenza vaccines demonstrate, the elderly are the least likely to seroconvert or generate high titers of neutralizing antibodies after immunization (Goodwin et al. 2006).

5.1   Live-Attenuated and Chimeric WNV Strains

Live-attenuated WNV vaccine strains are inherently less pathogenic yet can replicate and elicit both humoral and cellular immune responses in a manner akin to infection with virulent WNV strains. The WNV-25 strain was developed by serial passage in mosquito cells until mutations accumulated that reduced neuroinvasiveness greater than 10^6-fold (Chambers et al. 1998; Halevy et al. 1994). Mice infected with WNV-25 seroconvert as efficiently as those infected with parental virus

(Halevy et al. 1994), and immunization protected geese against lethal challenge with a virulent WNV isolate (Lustig et al. 2000). Analogously, mutation of a single amino acid residue in the NS2A gene (A30P) converted an inherently less virulent Australian subtype of WNV, Kunjin virus (KUNV), to a highly attenuated strain (Liu et al. 2006). This defect was associated with attenuation of viral antagonism of type I IFN responses. Intraperitoneal infection with A30P-KUN elicited antibody responses similar to the parent virus and significantly protected mice from both peripheral and CNS challenges with a virulent WNV strain. However, the introduction of additional attenuating mutations may be necessary to reduce the risk of parental virus reversion.

Live-attenuated WNV vaccine strains have also been generated by chimerization, which may further reduce the likelihood of viral reversion (Murphy et al. 2004). Transplantation of the prM and E structural genes of WNV into the infectious clone backbone of the 17D YFV vaccine strain, PDK-53 DENV-2 vaccine strain, or a Caribbean strain of DENV-4 has produced candidate chimeric WNV vaccine strains that induce specific neutralizing antibody responses (Arroyo et al. 2001; Huang et al. 2005; Monath 2001; Pletnev et al. 2002). Generation of these chimeric vaccine strains on virus backgrounds that are already attenuated limits the neuroinvasiveness and neurovirulence of these strains (Arroyo et al. 2001; Huang et al. 2005). Additionally, the process of chimerization independently attenuates *Flavivirus* virulence (Seligman et al. 2004). To further minimize the risk of reversion or toxicity, attenuating mutations have been introduced into chimeric WNV-YFV and WNV-DENV-4 vaccine strains (Arroyo et al. 2001; Pletnev et al. 2002; Pletnev et al. 2006). Despite all of the genetic changes, chimeric WNV vaccines still elicit effective neutralizing antibody responses. Immunization with a single dose of chimeric WNV-YFV virus induced neutralizing and complement-fixing antibodies in rodents and nonhuman primates and completely protected animals from a lethal challenge of virulent WNV (Arroyo et al. 2001; Huang et al. 2005; Pletnev et al. 2002; Tesh et al. 2002). In clinical trials in humans, administration of WNV-YFV chimeric virus to healthy volunteers resulted in transient viremia and a low incidence of flu-like symptoms. All subjects developed neutralizing antibodies and most developed antiviral T cell responses to WNV (Monath et al. 2006).

5.2   Formalin-Inactivated WNV Vaccines

Inactivated whole virus vaccines have been developed against WNV, and are currently in use in veterinary practice (Ng et al. 2003). Administration of one or two doses of formalin-treated WNV vaccine to geese resulted in 42% and 89% protection, respectively, at 3 weeks post-immunization (Malkinson et al. 2001). Similarly, two doses of inactivated WNV completely protected hamsters from lethal WNV challenge and elicited the development of neutralizing WNV-specific antibodies (Tesh et al. 2002). Vaccination of horses with inactivated WNV also stimulated long-lived neutralizing antibody responses within 14 days of vaccination (Ng et al. 2003).

Although killed WNV vaccines could be used to vaccinate the immunocompromised, their utility may be limited by their decreased immunogenicity. Administration of multiple vaccine doses may be required to elicit a protective immune response. Furthermore, it is unclear how durable immunity will be, as relatively low levels of neutralizing and complement-fixing antibodies were detected in hamsters 1 month after initial immunization (Tesh et al. 2002).

5.3   Subunit Vaccines, Nonreplicating Particles, and DNA Vaccines

Vaccination of animals with recombinant WNV proteins or subviral (prM-E) particles also induces neutralizing antibody responses. Repeated immunization of mice or hamsters with purified, recombinant WNV E protein resulted in the development of complement-fixing and neutralizing anti-WNV antibodies that were protective against lethal WNV challenge at least 1 year after vaccination (Ledizet et al. 2005; Wang et al. 2001; Watts et al. 2006). Neutralizing antibody responses to recombinant E protein immunization in horses appeared to be more potent and durable than that observed following vaccination with inactivated virus (Ledizet et al. 2005). Co-expression of prM and E protein alone can induce the formation of *Flavivirus* subviral particles that are immunogenic in mice (Konishi et al. 1992; Kroeger et al. 2002). Repeated immunization of mice with WNV subviral particles induced neutralizing antibody responses and protected mice against lethal WNV challenge, although a transient viremia was observed after challenge (Qiao et al. 2004). Similarly, DNA-based immunization with a plasmid encoding prM and E stimulated robust humoral immune responses against WNV (reviewed in Chang et al. 2001). A single dose of plasmid DNA encoding WNV prM and E (Davis et al. 2001) or C proteins (Yang et al. 2001) protected horses and outbred mice from a lethal dose of WNV. Because recombinant protein, subviral particles, or DNA plasmid vaccines are not infectious, they may be safely administered to the elderly and immunocompromised. Additionally, these vaccines may exhibit improved specificity because they direct the immune response to a few viral antigens. Nonetheless, multiple doses may be required to stimulate effective humoral responses, and, similar to inactivated virus preparations, it is unclear how durable the protective immune response will be to a nonreplicating protein antigen.

5.4   Other Viral Vectors

Vaccination with WNV antigens by transgenic expression from other viral vectors has also been tested. The Schwarz strain of measles virus was engineered to express WNV E protein; immunization induced potent neutralizing antibody responses in mice equivalent to that seen after WNV infection (Despres et al. 2005). Notably, immunization with the recombinant measles virus vector completely protected the highly susceptible IFN α/β receptor-deficient mice from lethal WNV challenge, and serum from vaccinated mice when passively transferred to BALB/c mice conferred protection. Analogously, immunization with a recombinant canarypox virus encoding WNV prM-E genes induced neutralizing antibodies in cats, dog, and horses, and prevented viremia upon challenge with WNV-infected mosquitoes (Karaca et al. 2005; Minke et al. 2004). Lentiviruses that encoded WNV E protein also induced protective neutralizing antibody responses in mice (Iglesias et al. 2006). Although these heterologous virus expression systems have been previously characterized for safety and efficacy, they are still infectious and their use in the elderly or immunocompromised may therefore be limited.

6   Antibody-Based Therapeutics Against WNV

Although antibody has been utilized as a therapeutic against several viral infections (Sawyer 2000; Zeitlin et al. 1999), it has not been used extensively against *Flavivirus* infections in humans, with the exception of its prophylactic use against TBE virus (Kreil et al. 1997; Roehrig et al. 2001). In theory, antibody-dependent enhancement of infection could complicate the therapeutic administration of antibodies in patients. As mentioned, despite its extensive characterization in vitro, its significance in vivo with WNV remains uncertain. Antibody-dependent enhancement has been described after passive acquisition of antibodies against yellow fever and Langat encephalitis viruses (Barrett et al. 1986; Gould et al. 1987, 1989; Wallace et al. 2003). However, it was not observed after transfer of monoclonal or polyclonal antibodies against Japanese encephalitis virus (Kimura-Kuroda et al. 1988) or TBE virus (Kreil et al. 1997).

Recent studies indicate that passive administration of WNV-specific antibodies is protective. Adverse effects related to immune enhancement in rodent models have not been observed. Transfer of immune serum prior to WNV infection protected mice from lethality, and no increased mortality was observed at subneutralizing doses (Diamond et al. 2003a). Similarly, passive administration of immune serum (Tesh et al. 2002) or anti-E antiserum (Wang et al. 2001) protected hamsters and mice against lethal WNV infection. Several groups have demonstrated that immune human γ-globulin or neutralizing mAbs can prevent or treat neuroinvasive WNV infection (Ben-Nathan et al. 2003; Engle et al. 2003; Gould et al. 2005; Julander et al. 2005; Morrey et al. 2006; Oliphant et al. 2005, 2006; Throsby et al. 2006). Therapeutic intervention even 5 days after infection reduced WNV mortality; this time point is significant because WNV spreads to the brain and spinal cord between days 4 and 5 (Morrey et al. 2006; Oliphant et al. 2005). However, administration of neutralizing antibody at day 6 after infection did not enhance survival in mice, although average survival time was increased. Apparently, there is a window for therapeutic intervention with antibodies, and efficacy may be limited once significant neurological injury occurs.

Small numbers of human patients have received antibody therapy against WNV infection. As vaccination may not provide protective humoral immunity to the populations most at risk for severe WNV disease, immunoprophylaxis and immunotherapy with neutralizing anti-WNV mAbs may serve as an effective intervention in the elderly and immunocompromised. Case reports have documented clinical improvement in humans with neurological WNV disease following treatment with neutralizing WNV-immune human γ-globulin (Hamdan et al. 2002; Shimoni et al. 2001). Because it is obtained from pooled donors, WNV-immune human γ-globulin has relatively modest neutralizing activity (Engle et al. 2003). As it requires large volume administration, WNV immune globulin may adversely affect patients with cardiac or renal disease. Moreover, because it is a pooled blood product, it also has an inherent risk of transmitting known and unknown infectious agents.

To overcome these limitations, humanized and human mAbs or MAb fragments with therapeutic activity against WNV infection have been isolated or engineered (Gould et al. 2005; Oliphant et al. 2005; Throsby et al. 2006). These mAbs or antibody fragments have been well characterized in vitro and in vivo for neutralizing activity and mechanism of action. Molecular cloning of human or humanized neutralizing mAbs offers another advantage, as the potency of these recombinant mAbs may be improved by introducing mutations that increase the strength of the antibody–antigen interaction. The use of a combination of human or humanized antibodies that bind distinct epitopes and neutralize by independent mechanisms could overcome the potential risk of rapidly selecting escape variants in vivo.

7 Conclusions

Over the last few years, significant advances have been made in our understanding of the molecular and structural basis of antibody-mediated neutralization of WNV. Despite this, many questions remain unanswered: (1) what is the mechanism of inhibition for the DI- and DII-specific mAb neutralizing mAbs that localize outside of the fusion loop; (2) what role do anti-prM and anti-M antibodies have in neutralization and protection; (3) how many antibody molecules must bind the virion to neutralize infection; (4) how important is complement activation to the activity of strongly neutralizing mAbs in vivo; (5) which E protein epitopes are immunodominant following natural WNV infection in humans and other animals; (6) why are antibodies against poorly exposed epitopes still protective in vitro and in vivo? Ultimately, answering these questions will inform the development of novel antibody-based therapeutics and vaccines against WNV and other flaviviruses that target specific epitopes.

Acknowledgements The authors thank T. Pierson, G. Nybakken and T. Oliphant for help with figure preparation and critical comments. This work was supported by Pediatric Dengue Vaccine Initiative, and by NIH grants AI53870 and AI061373 (M.S.D.), and U54 AI057160 to the Midwest Regional Center of Excellence for Biodefense and Emerging Infectious Diseases Research.

References

Allison SL, Schalich J, Stiasny K, Mandl CW, Kunz C, Heinz FX (1995) Oligomeric rearrangement of tick-borne encephalitis virus envelope proteins induced by an acidic pH. J Virol 69:695–700

Allison SL, Schalich J, Stiasny K, Mandl CW, Heinz FX (2001) Mutational evidence for an internal fusion peptide in flavivirus envelope protein E. J Virol 75:4268–4275

Anderson JF, Rahal JJ (2002) Efficacy of interferon alpha-2b and ribavirin against West Nile virus in vitro. Emerg Infect Dis 8:107–108

Arroyo J, Miller CA, Catalan J, Monath TP (2001) Yellow fever vector live-virus vaccines: West Nile virus vaccine development. Trends Mol Med 7:350–354

Barrett AD, Gould EA (1986) Antibody-mediated early death in vivo after infection with yellow fever virus. J Gen Virol 67:2539–2542

Beasley DW, Aaskov JG (2001) Epitopes on the dengue 1 virus envelope protein recognized by neutralizing IgM monoclonal antibodies. Virology 279:447–458

Beasley DW, Barrett AD (2002a) Identification of neutralizing epitopes within structural domain III of the West Nile virus envelope protein. J Virol 76:13097–13100

Beasley DW, Li L, Suderman MT, Barrett AD (2002b) Mouse neuroinvasive phenotype of West Nile virus strains varies depending upon virus genotype. Virology 296:17–23

Beasley DW, Whiteman MC, Zhang S, Huang CY, Schneider BS, Smith DR, Gromowski GD, Higgs S, Kinney RM, Barrett AD (2005) Envelope protein glycosylation status influences mouse neuroinvasion phenotype of genetic lineage 1 West Nile virus strains. J Virol 79:8339–8347

Ben-Nathan D, Lustig S, Tam G, Robinzon S, Segal S, Rager-Zisman B (2003) Prophylactic and therapeutic efficacy of human intravenous immunoglobulin in treating West Nile virus infection in mice. J Infect Dis 188:5–12

Berry DM, Almeida JD (1968) The morphological and biological effects of various antisera on avian infectious bronchitis virus. J Gen Virol 3:97–102

Berthet FX, Zeller HG, Drouet MT, Rauzier J, Digoutte JP, Deubel V (1997) Extensive nucleotide changes and deletions within the envelope glycoprotein gene of Euro-African West Nile viruses. J Gen Virol 78:2293–2297

Best SM, Morris KL, Shannon JG, Robertson SJ, Mitzel DN, Park GS, Boer E, Wolfinbarger JB, Bloom ME (2005) Inhibition of interferon-stimulated JAK-STAT signaling by a tick-borne flavivirus and identification of NS5 as an interferon antagonist. J Virol 79:12828–12839

Bhardwaj S, Holbrook M, Shope RE, Barrett AD, Watowich SJ (2001) Biophysical characterization and vector-specific antagonist activity of domain III of the tick-borne flavivirus envelope protein. J Virol 75:4002–4007

Brandriss MW, Schlesinger JJ, Walsh EE, Briselli M (1986) Lethal 17D yellow fever encephalitis in mice. I. Passive protection by monoclonal antibodies to the envelope proteins of 17D yellow fever and dengue 2 viruses. J Gen Virol 67:229–234

Brault AC, Langevin SA, Bowen RA, Panella NA, Biggerstaff BJ, Miller BR, Nicholas K (2004) Differential virulence of West Nile strains for American crows. Emerg Infect Dis 10:2161–2168

Bressanelli S, Stiasny K, Allison SL, Stura EA, Duquerroy S, Lescar J, Heinz FX, Rey FA (2004) Structure of a flavivirus envelope glycoprotein in its low-pH-induced membrane fusion conformation. EMBO J 23:728–738

Brinton MA (2002) The molecular biology of West Nile virus: a new invader of the western hemisphere. Annu Rev Microbiol 56:371–402

Burke DS, Monath TP (2001) Flaviviruses. In: Knipe DM, Howley PM (eds) Fields virology. Lippincott Williams & Wilkins, Philadelphia, pp 1043–1125

Burton DR, Saphire EO, Parren PW (2001) A model for neutralization of viruses based on antibody coating of the virion surface. Curr Top Microbiol Immunol 260:109–143

Busch MP, Caglioti S, Robertson EF, McAuley JD, Tobler LH, Kamel H, Linnen JM, Shyamala V, Tomasulo P, Kleinman SH (2005) Screening the blood supply for West Nile virus RNA by nucleic acid amplification testing. N Engl J Med 353:460–467

Butrapet S, Kimura-Kuroda J, Zhou DS, Yasui K (1998) Neutralizing mechanism of a monoclonal antibody against Japanese encephalitis virus glycoprotein E. Am J Trop Med Hyg 58:389–398

Byrne SN, Halliday GM, Johnston LJ, King NJ (2001) Interleukin-1beta but not tumor necrosis factor is involved in West Nile virus-induced Langerhans cell migration from the skin in C57BL/6 mice. J Invest Dermatol 117:702–709

Cardosa MJ, Porterfield JS, Gordon S (1983) Complement receptor mediates enhanced flavivirus replication in macrophages. J Exp Med 158:258–263

Cardosa MJ, Gordon S, Hirsch S, Springer TA, Porterfield JS (1986) Interaction of West Nile virus with primary murine macrophages: role of cell activation and receptors for antibody and complement. J Virol 57:952–959

Cecilia D, Gould EA (1991) Nucleotide changes responsible for loss of neuroinvasiveness in Japanese encephalitis virus neutralization-resistant mutants. Virology 181:70–77

Chambers TJ, Diamond MS (2003) Pathogenesis of flavivirus encephalitis. In: Chambers TJ, Monath TP (eds) The flaviviruses: current molecular aspects of evolution, biology, and disease prevention. Academic Press, San Diego, pp 273–342

Chambers TJ, Halevy M, Nestorowicz A, Rice CM, Lustig S (1998) West Nile virus envelope proteins: nucleotide sequence analysis of strains differing in mouse neuroinvasiveness. J Gen Virol 79:2375–2380

Chang GJ, Davis BS, Hunt AR, Holmes DA, Kuno G (2001) Flavivirus DNA vaccines: current status and potential. Ann N Y Acad Sci 951:272–285

Choi KS, Nah JJ, Ko, YJ, Kim YJ, Joo YS (2007) The DE loop of the domain III of the envelope protein appears to be associated with West Nile virus neutralization. Virus Res 123:216–218

Chu JJ, Ng ML (2004a) Interaction of West Nile virus with alpha v beta 3 integrin mediates virus entry into cells. J Biol Chem 279:54533–54541

Chu JJ, Ng ML (2004b) Infectious entry of West Nile virus occurs through a clathrin-mediated endocytic pathway. J Virol 78:10543–10555

Chu JJ, Rajamanonmani R, Li J, Bhuvanakantham R, Lescar J, Ng ML (2005) Inhibition of West Nile virus entry by using a recombinant domain III from the envelope glycoprotein. J Gen Virol 86:405–412

Chung KM, Liszewski MK, Nybakken G, Davis AE, Townsend RR, Fremont DH, Atkinson JP, Diamond MS (2006a) West Nile virus non-structural protein NS1 inhibits complement activation by binding the regulatory protein factor H. Proc Natl Acad Sci USA 103:19111–19116

Chung KM, Nybakken GE, Thompson BS, Engle MJ, Marri A, Fremont DH, Diamond MS (2006b) Antibodies against West Nile virus non-structural (NS)-1 protein prevent lethal infection through Fc gamma receptor-dependent and independent mechanisms. J Virol 80:1340–1351

Chung KM, Thompson BS, Fremont DH, Diamond MS (2007) Antibody recognition of cell surface-associated NS1 triggers Fc-gamma receptor mediated phagocytosis and clearance of West Nile virus infected cells. J Virol. In press.

Churdboonchart V, Bhamarapravati N, Peampramprecha S, Sirinavin S (1991) Antibodies against dengue viral proteins in primary and secondary dengue hemorrhagic fever. Am J Trop Med Hyg 44:481–493

Cooper NR, Jensen FC, Welsh RM Jr, Oldstone MB (1976) Lysis of RNA tumor viruses by human serum: direct antibody-independent triggering of the classical complement pathway. J Exp Med 144:970–984

Crill WD, Chang GJ (2004) Localization and characterization of flavivirus envelope glycoprotein cross-reactive epitopes. J Virol 78:13975–13986

Crill WD, Roehrig JT (2001) Monoclonal antibodies that bind to domain III of dengue virus E glycoprotein are the most efficient blockers of virus adsorption to Vero cells. J Virol 75:7769–7773

Davis BS, Chang GJ, Cropp B, Roehrig JT, Martin DA, Mitchell CJ, Bowen R, Bunning ML (2001) West Nile virus recombinant DNA vaccine protects mouse and horse from virus challenge and expresses in vitro a noninfectious recombinant antigen that can be used in enzyme-linked immunosorbent assays. J Virol 75:4040–4047

Davis CW, Mattei LM, Nguyen HY, Doms RW, Pierson TC (2006a) The location of N-linked glycans on West Nile virions controls their interactions with CD209. J Biol Chem 281:37183–37194

Davis CW, Nguyen HY, Hanna SL, Sanchez MD, Doms RW, Pierson TC (2006b) West Nile virus discriminates between DC-SIGN and DC-SIGNR for cellular attachment and infection. J Virol 80:1290–1301

Davis LE, DeBiasi R, Goade DE, Haaland KY, Harrington JA, Harnar JB, Pergam SA, King MK, DeMasters BK, Tyler KL (2006c) West Nile virus neuroinvasive disease. Ann Neurol 60:286–300

Deardorff E, Estrada-Franco J, Brault AC, Navarro-Lopez R, Campomanes-Cortes A, Paz-Ramirez P, Solis-Hernandez M, Ramey WN, Davis CT, Beasley DW, Tesh RB, Barrett AD, Weaver SC (2006) Introductions of West Nile virus strains to Mexico. Emerg Infect Dis 12:314–318

den Haan JM, Bevan MJ (2002) Constitutive versus activation-dependent cross-presentation of immune complexes by CD8(+) and CD8(−) dendritic cells in vivo. J Exp Med 196:817–827

Despres P, Combredet C, Frenkiel MP, Lorin C, Brahic M, Tangy F (2005) Live measles vaccine expressing the secreted form of the West Nile virus envelope glycoprotein protects against West Nile virus encephalitis. J Infect Dis 191:207–214

Diamond MS, Shrestha B, Marri A, Mahan D, Engle M (2003a) B cells and antibody play critical roles in the immediate defense of disseminated infection by West Nile encephalitis virus. J Virol 77:2578–2586

Diamond MS, Shrestha B, Mehlhop E, Sitati E, Engle M (2003b) Innate and adaptive immune responses determine protection against disseminated infection by West Nile encephalitis virus. Viral Immunol 16:259–278

Diamond MS, Sitati E, Friend L, Shrestha B, Higgs S, Engle M (2003c) Induced IgM protects against lethal West Nile virus infection. J Exp Med 198:1–11

Dimmock NJ (1993) Neutralization of animal viruses. Curr Top Microbiol Immunol 183:1–149

Edwards JP, Zhang X, Frauwirth KA, Mosser DM (2006) Biochemical and functional characterization of three activated macrophage populations. J Leukoc Biol 80:1298–1307

Egloff MP, Benarroch D, Selisko B, Romette JL, Canard B (2002) An RNA cap (nucleoside-2′-O-)-methyltransferase in the flavivirus RNA polymerase NS5: crystal structure and functional characterization. EMBO J 21:2757–2768

Elshuber S, Allison SL, Heinz FX, Mandl CW (2003) Cleavage of protein prM is necessary for infection of BHK-21 cells by tick-borne encephalitis virus. J Gen Virol 84:183–191

Engle M, Diamond MS (2003) Antibody prophylaxis and therapy against West Nile virus infection in wild type and immunodeficient mice. J Virol 77:12941–12949

Fredericksen BL, Smith M, Katze MG, Shi PY, Gale M (2004) The host response to West Nile virus infection limits spread through the activation of the interferon regulatory factor 3 pathway. J Virol 78:7737–7747

Gollins S, Porterfield J (1984) Flavivirus infection enhancement in macrophages: radioactive and biological studies on the effect of antibody and viral fate. J Gen Virol 65:1261–1272

Gollins S, Porterfield J (1986a) The uncoating and infectivity of the flavivirus West Nile on interaction with cells: effects of pH and ammonium chloride. J Gen Virol 67:1941–1950

Gollins SW, Porterfield JS (1985) Flavivirus infection enhancement in macrophages: an electron microscopic study of viral entry. J Gen Virol 66:1969–1982

Gollins SW, Porterfield JS (1986b) A new mechanism for the neutralization of enveloped viruses by antiviral antibody. Nature 321:244–246

Goodwin K, Viboud C, Simonsen L (2006) Antibody response to influenza vaccination in the elderly: a quantitative review. Vaccine 24:1159–1169

Gould EA, Buckley A (1989) Antibody-dependent enhancement of yellow fever and Japanese encephalitis virus neurovirulence. J Gen Virol 70:1605–1608

Gould EA, Buckley A, Groeger BK, Cane PA, Doenhoff M (1987) Immune enhancement of yellow fever virus neurovirulence for mice: studies of mechanisms involved. J Gen Virol 68:3105–3112

Gould LH, Sui J, Foellmer H, Oliphant T, Wang T, Ledizet M, Murakami A, Noonan K, Lambeth C, Kar K, Anderson JF, de Silva AM, Diamond MS, Koski RA, Marasco WA, Fikrig E (2005) Protective and therapeutic capacity of human single chain Fv-Fc fusion proteins against West Nile virus. J Virol 79:14606–14613

Griffin DE, Ubol S, Despres P, Kimura T, Byrnes A (2001) Role of antibodies in controlling alphavirus infection of neurons. Curr Top Microbiol Immunol 260:191–200

Guirakhoo F, Heinz FX, Kunz C (1989) Epitope model of tick-borne encephalitis virus envelope glycoprotein E: analysis of structural properties, role of carbohydrate side chain, and conformational changes occurring at acidic pH. Virology 169:90–99

Guirakhoo F, Bolin RA, Roehrig JT (1992) The Murray Valley encephalitis virus prM protein confers acid resistance to virus particles and alters the expression of epitopes within the R2 domain of E glycoprotein. Virology 191:921–931

Halevy M, Akov Y, Ben-Nathan D, Kobiler D, Lachmi B, Lustig S (1994) Loss of active neuroinvasiveness in attenuated strains of West Nile virus: pathogenicity in immunocompetent and SCID mice. Arch Virol 137:355–370

Halstead SB (1988) Pathogenesis of dengue: challenges to molecular biology. Science 239:476–481

Halstead SB (1994) Antibody-dependent enhancement of infection: a mechanism for indirect virus entry into cells. In: Cellular receptors for animal viruses. Cold Spring Harbor Laboratory Press, Cold Spring Harbor, pp 493–516

Halstead SB, O'Rourke EJ (1977) Antibody-enhanced dengue virus infection in primate leukocytes. Nature 265:739–741

Halstead SB, Venkateshan CN, Gentry MK, Larsen LK (1984) Heterogeneity of infection enhancement of dengue 2 strains by monoclonal antibodies. J Immunol 132:1529–1532

Hamano Y, Arase H, Saisho H, Saito T (2000) Immune complex and Fc receptor-mediated augmentation of antigen presentation for in vivo Th cell responses. J Immunol 164:6113–6119

Hamdan A, Green P, Mendelson E, Kramer MR, Pitlik S, Weinberger M (2002) Possible benefit of intravenous immunoglobulin therapy in a lung transplant recipient with West Nile virus encephalitis. Transpl Infect Dis 4:160–162

Hanna SL, Pierson TC, Sanchez MD, Ahmed AA, Murtadha MM, Doms RW (2005) N-linked glycosylation of West Nile virus envelope proteins influences particle assembly and infectivity. J Virol 79:13262–13274

Hayes EB, Komar N, Nasci RS, Montgomery SP, O'Leary DR, Campbell GL (2005a) Epidemiology and transmission dynamics of West Nile virus disease. Emerg Infect Dis 11:1167–1173

Hayes EB, Sejvar JJ, Zaki SR, Lanciotti RS, Bode AV, Campbell GL (2005b) Virology, pathology, and clinical manifestations of West Nile virus disease. Emerg Infect Dis 11:1174–1179

Heinz F, Auer G, Stiasny K, Holzmann H, Mandl C, Guirakhoo F, Kunz C (1994a) The interactions of the flavivirus envelope proteins: implications for virus entry and release. Arch Virol 9:339–348

Heinz F, Stiasny K, Puschner-Auer G, Holzmann H, Allison S, Mandl C, Kunz C (1994b) Structural changes and functional control of the tick-borne encephalitis virus glycoprotein E by the heterodimeric association with the protein prM. Virology 198:109–117

Heinz FX, Berger R, Tuma W, Kunz C (1983) A topological and functional model of epitopes on the structural glycoprotein of tick-borne encephalitis virus defined by monoclonal antibodies. Virology 126:525–537

Henchal EA, Henchal LS, Thaisomboonsuk BK (1987) Topological mapping of unique epitopes on the dengue-2 virus NS1 protein using monoclonal antibodies. J Gen Virol 68:845–851

Henchal EA, Henchal LS, Schlesinger JJ (1988) Synergistic interactions of anti-NS1 monoclonal antibodies protect passively immunized mice from lethal challenge with dengue 2 virus. J Gen Virol 69:2101–2107

Huang CY, Silengo SJ, Whiteman MC, Kinney RM (2005) Chimeric dengue 2 PDK-53/West Nile NY99 viruses retain the phenotypic attenuation markers of the candidate PDK-53 vaccine virus and protect mice against lethal challenge with West Nile virus. J Virol 79:7300–7310

Hubalek Z, Halouzka J (1999a) West Nile fever—a reemerging mosquito-borne viral disease in Europe. Emerg Infect Dis 5:643–650

Hubalek Z, Halouzka J, Juricova Z (1999b) West Nile fever in Czechland. Emerg Infect Dis 5:594–595

Huhn GD, Austin C, Langkop C, Kelly K, Lucht R, Lampman R, Novak R, Haramis L, Boker R, Smith S, Chudoba M, Gerber S, Conover C, Dworkin MS (2005) The emergence of West Nile virus during a large outbreak in Illinois in 2002. Am J Trop Med Hyg 72:768–776

Hung SL, Lee PL, Chen HW, Chen LK, Kao CL, King CC (1999) Analysis of the steps involved in Dengue virus entry into host cells. Virology 257:156–167

Hunsperger EA, Roehrig JT (2006) Temporal analyses of the neuropathogenesis of a West Nile virus infection in mice. J Neurovirol 12:129–139

Iglesias MC, Frenkiel MP, Mollier K, Souque P, Despres P, Charneau P (2006) A single immunization with a minute dose of a lentiviral vector-based vaccine is highly effective at eliciting protective humoral immunity against West Nile virus. J Gene Med 8:265–274

Johnston LJ, Halliday GM, King NJ (1996) Phenotypic changes in Langerhans' cells after infection with arboviruses: a role in the immune response to epidermally acquired viral infection? J Virol 70:4761–4766

Johnston LJ, Halliday GM, King NJ (2000) Langerhans cells migrate to local lymph nodes following cutaneous infection with an arbovirus. J Invest Dermatol 114:560–568

Julander JG, Winger QA, Olsen AL, Day CW, Sidwell RW, Morrey JD (2005) Treatment of West Nile virus-infected mice with reactive immunoglobulin reduces fetal titers and increases dam survival. Antiviral Res 65:79–85

Jupp PG (2001) The ecology of West Nile virus in South Africa and the occurrence of outbreaks in humans. Ann N Y Acad Sci 951:143–152

Kanai R, Kar K, Anthony K, Gould LH, Ledizet M, Fikrig E, Marasco WA, Koski RA, Modis Y (2006) Crystal structure of West Nile virus envelope glycoprotein reveals viral surface epitopes. J Virol 80:11000–11008

Karaca K, Bowen R, Austgen LE, Teehee M, Siger L, Grosenbaugh D, Loosemore L, Audonnet JC, Nordgren R, Minke JM (2005) Recombinant canarypox vectored West Nile virus (WNV) vaccine protects dogs and cats against a mosquito WNV challenge. Vaccine 23:3808–3813

Kauffman B, Nybakken G, Chipman PR, Zhang W, Fremont DH, Diamond MS, Kuhn RJ, Rossmann MG (2006) West Nile virus in complex with a neutralizing monoclonal antibody. Proc Natl Acad Sci USA 103:12400–12404

Keller BC, Fredericksen BL, Samuel MA, Mock RE, Mason PW, Diamond MS, Gale M Jr (2006) Resistance to alpha/beta interferon is a determinant of West Nile virus replication fitness and virulence. J Virol 80:9424–9434

Khromykh AA, Sedlak PL, Westaway EG (1999) trans-Complementation analysis of the flavivirus Kunjin ns5 gene reveals an essential role for translation of its N-terminal half in RNA replication. J Virol 73:9247–9255

Khromykh AA, Sedlak PL, Westaway EG (2000) cis- and trans-acting elements in flavivirus RNA replication. J Virol 74:3253–3263

Kile JC, Panella NA, Komar N, Chow CC, MacNeil A, Robbins B, Bunning ML (2005) Serologic survey of cats and dogs during an epidemic of West Nile virus infection in humans. J Am Vet Med Assoc 226:1349–1353

Kimura-Kuroda J, Yasui K (1983) Topographical analysis of antigenic determinants on envelope glycoprotein V3 (E) of Japanese encephalitis virus, using monoclonal antibodies. J Virol 45:124–132

Kimura-Kuroda J, Yasui K (1988) Protection of mice against Japanese encephalitis virus by passive administration with monoclonal antibodies. J Immunol 141:3606–3610

Kliks S (1990) Antibody-enhanced infection of monocytes as the pathogenetic mechanism for severe dengue illness. AIDS Res Hum Retroviruses 6:993–998

Kliks SC, Nisalak A, Brandt WE, Wahl L, Burke DS (1989) Antibody-dependent enhancement of dengue virus growth in human monocytes as a risk factor for dengue hemorrhagic fever. Am J Trop Med Hyg 40:444–451

Komar N, Clark GG (2006) West Nile virus activity in Latin America and the Caribbean. Rev Panam Salud Publica 19:112–117

Konishi E, Pincus S, Paoletti E, Shope RE, Burrage T, Mason PW (1992) Mice immunized with a subviral particle containing the Japanese encephalitis virus prM/M and E proteins are protected from lethal JEV infection. Virology 188:714–720

Kreil TR, Eibl MM (1997) Pre- and postexposure protection by passive immunoglobulin but no enhancement of infection with a flavivirus in a mouse model. J Virol 71:2921–2927

Kroeger MA, McMinn PC (2002) Murray Valley encephalitis virus recombinant subviral particles protect mice from lethal challenge with virulent wild-type virus. Arch Virol 147:1155–1172

Kuhn RJ, Zhang W, Rossmann MG, Pletnev SV, Corver J, Lenches E, Jones CT, Mukhopadhyay S, Chipman PR, Strauss EG, Baker TS, Strauss JH (2002) Structure of dengue virus: implications for flavivirus organization, maturation, and fusion. Cell 108:717–725

Lanciotti RS, Ebel GD, Deubel V, Kerst AJ, Murri S, Meyer R, Bowen M, McKinney N, Morrill WE, Crabtree MB, Kramer LD, Roehrig JT (2002) Complete genome sequences and phylogenetic analysis of West Nile virus strains isolated from the United States, Europe, and the Middle East. Virology 298:96–105

Ledizet M, Kar K, Foellmer HG, Wang T, Bushmich SL, Anderson JF, Fikrig E, Koski RA (2005) A recombinant envelope protein vaccine against West Nile virus. Vaccine 23:3915–3924

Levine B, Hardwick JM, Trapp BD, Crawford TO, Bollinger RC, Griffin DE (1991) Antibody-mediated clearance of alphavirus infection from neurons. Science 254:856–860

Lim JK, Glass WG, McDermott DH, Murphy PM (2006) CCR5: no longer a "good for nothing" gene—chemokine control of West Nile virus infection. Trends Immunol 27:308–312

Lin B, Parrish CR, Murray JM, Wright PJ (1994) Localization of a neutralizing epitope on the envelope protein of dengue virus type 2. Virology 202:885–890

Lin RJ, Chang BL, Yu HP, Liao CL, Lin YL (2006) Blocking of interferon-induced Jak-Stat signaling by Japanese encephalitis virus NS5 through a protein tyrosine phosphatase-mediated mechanism. J Virol 80:5908–5918

Lindenbach BD, Rice CM (1999) Genetic interaction of flavivirus nonstructural proteins NS1 and NS4A as a determinant of replicase function. J Virol 73:4611–4621

Lindenbach BD, Rice CM (2001) Flaviviridae: the viruses and their replication. In: Knipe DM, Howley PM (eds) Fields virology. Lippincott Williams & Wilkins, Philadelphia, pp 991–1041

Liu WJ, Chen HB, Khromykh AA (2003) Molecular and functional analyses of Kunjin virus infectious cDNA clones demonstrate the essential roles for NS2A in virus assembly and for a nonconservative residue in NS3 in RNA replication. J Virol 77:7804–7813

Liu WJ, Chen HB, Wang XJ, Huang H, Khromykh AA (2004) Analysis of adaptive mutations in Kunjin virus replicon RNA reveals a novel role for the flavivirus nonstructural protein NS2A in inhibition of beta interferon promoter-driven transcription. J Virol 78:12225–12235

Liu WJ, Wang XJ, Mokhonov VV, Shi PY, Randall R, Khromykh AA (2005) Inhibition of interferon signaling by the New York 99 strain and Kunjin subtype of West Nile virus involves blockage of STAT1 and STAT2 activation by nonstructural proteins. J Virol 79:1934–1942

Liu WJ, Wang XJ, Clark DC, Lobigs M, Hall RA, Khromykh AA (2006) A single amino acid substitution in the West Nile virus nonstructural protein NS2A disables its ability to inhibit alpha/beta interferon induction and attenuates virus virulence in mice. J Virol 80:2396–2404

Lucas M, Mashimo T, Frenkiel MP, Simon-Chazottes D, Montagutelli X, Ceccaldi PE, Guenet JL, Despres P (2003) Infection of mouse neurones by West Nile virus is modulated by the interferon-inducible 2′-5′ oligoadenylate synthetase 1b protein. Immunol Cell Biol 81:230–236

Lustig S, Olshevsky U, Ben-Nathan D, Lachmi BE, Malkinson M, Kobiler D, Halevy M (2000) A live attenuated West Nile virus strain as a potential veterinary vaccine. Viral Immunol 13:401–410

Mackenzie JM, Westaway EG (2001) Assembly and maturation of the flavivirus Kunjin virus appear to occur in the rough endoplasmic reticulum and along the secretory pathway, respectively. J Virol 75:10787–10799

Malkinson M, Banet C, Khinich Y, Samina I, Pokamunski S, Weisman Y (2001) Use of live and inactivated vaccines in the control of West Nile fever in domestic geese. Ann N Y Acad Sci 951:255–261

Mashimo T, Lucas M, Simon-Chazottes D, Frenkiel MP, Montagutelli X, Ceccaldi PE, Deubel V, Guenet JL, Despres P (2002) A nonsense mutation in the gene encoding 2′-5′-oligoadenylate synthetase/L1 isoform is associated with West Nile virus susceptibility in laboratory mice. Proc Natl Acad Sci USA 99:11311–11316

Mason PW, Dalrymple JM, Gentry MK, McCown JM, Hoke CH, Burke DS, Fournier MJ, Mason TL (1989) Molecular characterization of a neutralizing domain of the Japanese encephalitis virus structural glycoprotein. J Gen Virol 70:2037–2049

Megret F, Hugnot JP, Falconar A, Gentry MK, Morens DM, Murray JM, Schlesinger JJ, Wright PJ, Young P, Van Regenmortel MH, et al (1992) Use of recombinant fusion proteins and monoclonal antibodies to define linear and discontinuous antigenic sites on the dengue virus envelope glycoprotein. Virology 187:480–491

Mehlhop E, Diamond MS (2006) Protective immune responses against West Nile virus are primed by distinct complement activation pathways. J Exp Med 203:1371–1381

Mehlhop E, Whitby K, Oliphant T, Marri A, Engle M, Diamond MS (2005) Complement activation is required for the induction of a protective antibody response against West Nile virus infection. J Virol 79:7466–7477

Miller DL, Mauel MJ, Baldwin C, Burtle G, Ingram D, Hines ME, 2nd Frazier KS (2003) West Nile virus in farmed alligators. Emerg Infect Dis 9:794–799

Minke JM, Siger L, Karaca K, Austgen L, Gordy P, Bowen R, Renshaw RW, Loosmore S, Audonnet JC, Nordgren B (2004) Recombinant canarypoxvirus vaccine carrying the prM/E genes of West Nile virus protects horses against a West Nile virus-mosquito challenge. Arch Virol Suppl 221–230

Modis Y, Ogata S, Clements D, Harrison SC (2003) A ligand-binding pocket in the dengue virus envelope glycoprotein. Proc Natl Acad Sci USA 100:6986–6991

Modis Y, Ogata S, Clements D, Harrison SC (2004) Structure of the dengue virus envelope protein after membrane fusion. Nature 427:313–319

Modis Y, Ogata S, Clements D, Harrison SC (2005) Variable surface epitopes in the crystal structure of dengue virus type 3 envelope glycoprotein. J Virol 79:1223–1231

Monath TP (2001) Prospects for development of a vaccine against the West Nile virus. Ann NY Acad Sci 951:1–12

Monath TP, Liu J, Kanesa-Thasan N, Myers GA, Nichols R, Deary A, McCarthy K, Johnson C, Ermak T, Shin S, Arroyo J, Guirakhoo F, Kennedy JS, Ennis FA, Green S, Bedford P (2006) A live, attenuated recombinant West Nile virus vaccine. Proc Natl Acad Sci USA 103:6694–6699

Morens DM (1994) Antibody-dependent of enhancement of infection and the pathogenesis of viral disease. Clin Infect Dis 19:500–512

Morrey JD, Siddharthan V, Olsen AL, Roper GY, Wang H, Baldwin TJ, Koenig S, Johnson S, Nordstrom JL, Diamond MS (2006) Humanized monoclonal antibody against West Nile virus E protein administered after neuronal infection protects against lethal encephalitis in hamsters. J Infect Dis 194:1300–1308

Mostashari F, Bunning ML, Kitsutani PT, Singer DA, Nash D, Cooper MJ, Katz N, Liljebjelke KA, Biggerstaff BJ, Fine AD, Layton MC, Mullin SM, Johnson AJ, Martin DA, Hayes EB, Campbell GL (2001) Epidemic West Nile encephalitis, New York, 1999: results of a household-based seroepidemiological survey. Lancet 358:261–264

Mukhopadhyay S, Kim BS, Chipman PR, Rossmann MG, Kuhn RJ (2003) Structure of West Nile virus. Science 302:248

Munoz-Jordan JL, Sanchez-Burgos GG, Laurent-Rolle M, Garcia-Sastre A (2003) Inhibition of interferon signaling by dengue virus. Proc Natl Acad Sci USA 100:14333–14338

Munoz-Jordan JL, Laurent-Rolle M, Ashour J, Martinez-Sobrido L, Ashok M, Lipkin WI, Garcia-Sastre A (2005) Inhibition of alpha/beta interferon signaling by the NS4B protein of flaviviruses. J Virol 79:8004–8013

Murphy BR, Blaney JE Jr, Whitehead SS (2004) Arguments for live flavivirus vaccines. Lancet 364:499–500

Ng T, Hathaway D, Jennings N, Champ D, Chiang YW, Chu HJ (2003) Equine vaccine for West Nile virus. Dev Biol (Basel) 114:221–227

Nimmerjahn F, Ravetch JV (2006) Fcgamma receptors: old friends and new family members. Immunity 24:19–28

Nimmerjahn F, Bruhns P, Horiuchi K, Ravetch JV (2005) FcgammaRIV: a novel FcR with distinct IgG subclass specificity. Immunity 23:41–51

Nowak T, Farber PM, Wengler G (1989) Analyses of the terminal sequences of West Nile virus structural proteins and of the in vitro translation of these proteins allow the proposal of a complete scheme of the proteolytic cleavages involved in their synthesis. Virology 169:365–376

Nybakken G, Oliphant T, Johnson S, Burke S, Diamond MS, Fremont DH (2005) Structural basis for neutralization of a therapeutic antibody against West Nile virus. Nature 437:764–769

Nybakken GE, Nelson CA, Chen BR, Diamond MS, Fremont DH (2006) Crystal structure of the West Nile virus envelope glycoprotein. J Virol 80:11467–11474

Oliphant T, Engle M, Nybakken G, Doane C, Johnson S, Huang L, Gorlatov S, Mehlhop E, Marri A, Chung KM, Ebel GD, Kramer LD, Fremont DH, Diamond MS (2005) Development of a humanized monoclonal antibody with therapeutic potential against West Nile virus. Nat Med 11:522–530

Oliphant T, Nybakken G, Engle M, Xu, Q, Nelson CA, Sukupolvi-Petty S, Marri A, Lachmi B, Olshevsky U, Fremont DH, Pierson TC, Diamond MS (2006) Determinants of West Nile virus envelope protein domains I and II antibody recognition and neutralization. J Virol 80:12149–12159

Parren PW, Burton DR (2001) The antiviral activity of antibodies in vitro and in vivo. Adv Immunol 77:195–262

Peiris JS, Porterfield JS (1979) Antibody-mediated enhancement of flavivirus replication in macrophage-like cell lines. Nature 282:509–511

Peiris JS, Gordon S, Unkeless JC, Porterfield JS (1981) Monoclonal anti-Fc receptor IgG blocks antibody-dependent enhancement of viral replication in macrophages. Nature 289:189–191

Peiris JS, Porterfield JS, Roehrig JT (1982) Monoclonal antibodies against the flavivirus West Nile. J Gen Virol 58:283–289

Perelygin AA, Scherbik SV, Zhulin IB, Stockman BM, Li Y, Brinton MA (2002) Positional cloning of the murine flavivirus resistance gene. Proc Natl Acad Sci USA 99:9322–9327

Petersen LR, Marfin AA, Gubler DJ (2003) West Nile virus. JAMA 290:524–528

Pletnev AG, Putnak R, Speicher J, Wagar EJ, Vaughn DW (2002) West Nile virus/dengue type 4 virus chimeras that are reduced in neurovirulence and peripheral virulence without loss of immunogenicity or protective efficacy. Proc Natl Acad Sci USA 99:3036–3041

Pletnev AG, Swayne DE, Speicher J, Rumyantsev AA, Murphy BR (2006) Chimeric West Nile/dengue virus vaccine candidate: preclinical evaluation in mice, geese and monkeys for safety and immunogenicity. Vaccine 24:6392–6404

Qiao M, Ashok M, Bernard KA, Palacios G, Zhou ZH, Lipkin WI, Liang TJ (2004) Induction of sterilizing immunity against West Nile virus (WNV), by immunization with WNV-like particles produced in insect cells. J Infect Dis 190:2104–2108

Razumov IA, Kazachinskaia EI, Ternovoi VA, Protopopova EV, Galkina IV, Gromashevskii VL, Prilipov AG, Kachko AV, Ivanova AV, L'vov DK, Loktev VB (2005) Neutralizing monoclonal antibodies against Russian strain of the West Nile virus. Viral Immunol 18:558–568

Regnault A, Lankar D, Lacabanne V, Rodriguez A, Thery C, Rescigno M, Saito T, Verbeek S, Bonnerot C, Ricciardi-Castagnoli P, Amigorena S (1999) Fcgamma receptor-mediated induction of dendritic cell maturation and major histocompatibility complex class I-restricted antigen presentation after immune complex internalization. J Exp Med 189:371–380

Rey FA (2003) Dengue virus envelope glycoprotein structure: new insight into its interactions during viral entry. Proc Natl Acad Sci USA 100:6899–6901

Rey FA, Heinz FX, Mandl C, Kunz C, Harrison SC (1995) The envelope glycoprotein from tick-borne encephalitis virus at 2-angstrom resolution. Nature 375:291–298

Roehrig JT, Johnson AJ, Hunt AR, Bolin RA, Chu MC (1990) Antibodies to dengue 2 virus E-glycoprotein synthetic peptides identify antigenic conformation. Virology 177:668–675

Roehrig JT, Bolin RA, Kelly RG (1998) Monoclonal antibody mapping of the envelope glycoprotein of the dengue 2 virus, Jamaica. Virology 246:317–328

Roehrig JT, Staudinger LA, Hunt AR, Mathews JH, Blair CD (2001) Antibody prophylaxis and therapy for flaviviral encephalitis infections. Ann NY Acad Sci 951:286–297

Samuel MA, Diamond MS (2005) Type I IFN protects against lethal West Nile virus infection by restricting cellular tropism and enhancing neuronal survival. J Virol 79:13350–13361

Samuel MA, Diamond MS (2006a) Pathogenesis of West Nile virus infection: a balance between virulence, innate and adaptive immunity, and viral evasion. J Virol 80:9349–9360

Samuel MA, Whitby K, Keller BC, Marri A, Barchet W, Williams BR, Silverman RH, Gale M Jr, Diamond MS (2006b) PKR and RNAse L contribute to protection against lethal West Nile virus infection by controlling early viral spread in the periphery and replication in neurons. J Virol 80:7009–7019

Sanchez MD, Pierson TC, McAllister D, Hanna SL, Puffer BA, Valentine LE, Murtadha MM, Hoxie JA, Doms RW (2005) Characterization of neutralizing antibodies to West Nile virus. Virology 336:70–82

Sawyer LA (2000) Antibodies for the prevention and treatment of viral diseases. Antiviral Res 47:57–77

Scherbik SV, Paranjape JM, Stockman BM, Silverman RH, Brinton MA (2006) RNase L plays a role in the antiviral response to West Nile virus. J Virol 80:2987–2999

Schlesinger JJ, Chapman S (1995) Neutralizing F(ab')2 fragments of protective monoclonal antibodies to yellow fever virus (YF) envelope protein fail to protect mice against lethal YF encephalitis. J Gen Virol 76:217–220

Schlesinger JJ, Brandriss MW, Walsh EE (1985) Protection against 17D yellow fever encephalitis in mice by passive transfer of monoclonal antibodies to the nonstructural glycoprotein gp48 and by active immunization with gp48. J Immunol 135:2805–2809

Schlesinger JJ, Foltzer M, Chapman S (1993) The Fc portion of antibody to yellow fever virus NS1 is a determinant of protection against YF encephalitis in mice. Virology 192:132–141

Se-Thoe SY, Ling AE, Ng MM (2000) Alteration of virus entry mode: a neutralisation mechanism for Dengue-2 virus. J Med Virol 62:364–376

Sejvar JJ, Haddad MB, Tierney BC, Campbell GL, Marfin AA, Van Gerpen JA, Fleischauer A, Leis AA, Stokic DS, Petersen LR (2003) Neurologic manifestations and outcome of West Nile virus infection. JAMA 290:511–515

Sejvar JJ, Bode AV, Marfin AA, Campbell GL, Pape J, Biggerstaff BJ, Petersen LR (2006) West Nile Virus-associated flaccid paralysis outcome. Emerg Infect Dis 12:514–516

Seligman SJ, Gould EA (2004) Live flavivirus vaccines: reasons for caution. Lancet 363:2073–2075

Shimoni Z, Niven MJ, Pitlick S, Bulvik S (2001) Treatment of West Nile virus encephalitis with intravenous immunoglobulin. Emerg Infect Dis 7:759

Shirato K, Miyoshi H, Goto A, Ako Y, Ueki T, Kariwa H, Takashima I (2004) Viral envelope protein glycosylation is a molecular determinant of the nueroinvasiveness of the New York strain of West Nile virus. J Gen Virol 85:3637–3645

Shrestha B, Diamond MS (2004) The role of CD8+ T cells in the control of West Nile virus infection. J Virol 78:8312–8321

Shrestha B, Gottlieb DI, Diamond MS (2003) Infection and injury of neurons by West Nile encephalitis virus. J Virol 77:13203–13213

Shrestha B, Samuel MA, Diamond MS (2006a) CD8+ T cells require perforin to clear West Nile virus from infected neurons. J Virol 80:119–129

Shrestha B, Wang T, Samuel MA, Whitby K, Craft J, Fikrig E, Diamond MS (2006b) Gamma interferon plays a crucial early antiviral role in protection against West Nile virus infection. J Virol 80:5338–5348

Sitati E, Diamond MS (2006) CD4+ T Cell responses are required for clearance of West Nile virus from the central nervous system. J Virol 80:12060–12069

Sitati E, McCandless EE, Klein RS, Diamond MS (2007) CD40-CD40 ligand interactions promote trafficking of CD8+ T cells into the brain and protection against West Nile virus encephalitis. J Virol. In press.
Smithburn KC, Hughes TP, Burke AW, Paul JH (1940) A neurotropic virus isolated from the blood
of a native of Uganda. Am J Trop Med Hyg 20:471–492
Stadler K, Allison SL, Schalich J, Heinz FX (1997) Proteolytic activation of tick-borne encephalitis virus by furin. J Virol 71:8475–8481
Stiasny K, Kiermayr S, Holzmann H, Heinz FX (2006) Cryptic properties of a cluster of dominant
flavivirus cross-reactive antigenic sites. J Virol 80:9557–9568
Tesh RB, Arroyo J, Travassos Da Rosa AP, Guzman H, Xiao SY, Monath TP (2002) Efficacy of
killed virus vaccine, live attenuated chimeric virus vaccine, and passive immunization for prevention of West Nile virus encephalitis in hamster model. Emerg Infect Dis 8:1392–1397
Throsby M, Geuijen C, Goudsmit J, Bakker AQ, Korimbocus J, Kramer RA, Clijsters-van der Horst M,
de Jong M, Jongeneelen M, Thijsse S, Smit R, Visser TJ, Bijl N, Marissen WE, Loeb M, Kelvin
DJ, Preiser W, ter Meulen J, de Kruif J (2006) Isolation and characterization of human monoclonal antibodies from individuals infected with West Nile virus. J Virol 80:6982–6992
Tsai TF, Popovici F, Cernescu C, Campbell GL, Nedelcu NI (1998) West Nile encephalitis epi-
demic in southeastern Romania. Lancet 352:767–771
Valdes K, Alvarez M, Pupo M, Vazquez S, Rodriguez R, Guzman MG (2000) Human dengue
antibodies against structural and nonstructural proteins. Clin Diagn Lab Immunol 7:856–857
Vazquez S, Guzman MG, Guillen G, Chinea G, Perez AB, Pupo M, Rodriguez R, Reyes O, Garay
HE, Delgado I, Garcia G, Alvarez M (2002) Immune response to synthetic peptides of dengue
prM protein. Vaccine 20:1823–1830
Volk DE, Beasley DW, Kallick DA, Holbrook MR, Barrett AD, Gorenstein DG (2004) Solution
structure and antibody binding studies of the envelope protein domain III from the New York
strain of West Nile virus. J Biol Chem 279:38755–38761
Wallace MJ, Smith DW, Broom AK, Mackenzie JS, Hall RA, Shellam GR, McMinn PC (2003)
Antibody-dependent enhancement of Murray Valley encephalitis virus virulence in mice.
J Gen Virol 84:1723–1728
Wang T, Fikrig E (2004a) Immunity to West Nile virus. Curr Opin Immunol 16:519–523
Wang T, Anderson JF, Magnarelli LA, Wong SJ, Koski RA, Fikrig E (2001) Immunization of mice
against West Nile virus with recombinant envelope protein. J Immunol 167:5273–5277
Wang T, Scully E, Yin Z, Kim JH, Wang S, Yan J, Mamula M, Anderson JF, Craft J, Fikrig E
(2003a) IFN-γ-producing γδ T cells help control murine West Nile virus infection. J Immunol
171:2524–2531
Wang T, Town T, Alexopoulou L, Anderson JF, Fikrig E, Flavell RA (2004b) Toll-like receptor 3
mediates West Nile virus entry into the brain causing lethal encephalitis. Nat Med
10:1366–1373
Wang Y, Lobigs M, Lee E, Mullbacher A (2003b) CD8+ T cells mediate recovery and immunopa-
thology in West Nile virus encephalitis. J Virol 77:13323–13334
Wang Y, Lobigs M, Lee E, Mullbacher A (2004c) Exocytosis and Fas mediated cytolytic mechanisms exert protection from West Nile virus induced encephalitis in mice. Immunol Cell Biol
82:170–173
Watts DM, Tesh RB, Siirin M, Rosa AT, Newman PC, Clements DE, Ogata S, Coller BA, Weeks-Levy C,
Lieberman MM (2006) Efficacy and durability of a recombinant subunit West Nile vaccine candi-
date in protecting hamsters from West Nile encephalitis. Vaccine 25:2913–2918
Wengler G (1989) Cell-associated West Nile flavivirus is covered with E+pre-M protein heterodimers which are destroyed and reorganized by proteolytic cleavage during virus release.
J Virol 63:2521–2526
Winkler G, Randolph VB, Cleaves GR, Ryan TE, Stollar V (1988) Evidence that the mature form
of the flavivirus nonstructural protein NS1 is a dimer. Virology 162:187–196
Winkler G, Maxwell SE, Ruemmler C, Stollar V (1989) Newly synthesized dengue-2 virus nonstructural protein NS1 is a soluble protein but becomes partially hydrophobic and membrane-
associated after dimerization. Virology 171:302–305

Wong SJ, Boyle RH, Demarest VL, Woodmansee AN, Kramer LD, Li H, Drebot M, Koski RA, Fikrig E, Martin DA, Shi PY (2003) Immunoassay targeting nonstructural protein 5 to differentiate West Nile virus infection from dengue and St. Louis encephalitis virus infections and from flavivirus vaccination. J Clin Microbiol 41:4217–4223

Wu KP, Wu CW, Tsao YP, Kuo TW, Lou YC, Lin CW, Wu, SC, Cheng JW (2003) Structural basis of a flavivirus recognized by its neutralizing antibody: solution structure of the domain III of the Japanese Encephalitis virus envelope protein. J Biol Chem 278:46007–46013

Xiao SY, Guzman H, Zhang H, Travassos da Rosa AP, Tesh RB (2001) West Nile virus infection in the golden hamster (Mesocricetus auratus): a model for West Nile encephalitis. Emerg Infect Dis 7:714–721

Yang JS, Kim JJ, Hwang D, Choo AY, Dang K, Maguire H, Kudchodkar S, Ramanathan MP, Weiner DB (2001) Induction of potent Th1-type immune responses from a novel DNA vaccine for West Nile virus New York isolate (WNV-NY1999). J Infect Dis 184:809–816

Yusof R, Clum S, Wetzel M, Murthy HM, Padmanabhan R (2000) Purified NS2B/NS3 serine protease of dengue virus type 2 exhibits cofactor NS2B dependence for cleavage of substrates with dibasic amino acids in vitro. J Biol Chem 275:9963–9969

Zeitlin L, Cone RA, Whaley KJ (1999) Using monoclonal antibodies to prevent mucosal transmission of epidemic infectious diseases. Emerg Infect Dis 5:54–64

Zhang W, Chipman PR, Corver J, Johnson PR, Zhang Y, Mukhopadhyay S, Baker TS, Strauss JH, Rossmann MG, Kuhn RJ (2003a) Visualization of membrane protein domains by cryo-electron microscopy of dengue virus. Nat Struct Biol 10:907–912

Zhang Y, Corver J, Chipman PR, Zhang W, Pletnev SV, Sedlak D, Baker TS, Strauss JH, Kuhn RJ, Rossmann MG (2003b) Structures of immature flavivirus particles. EMBO J 22:2604–2613

Zhang Y, Zhang W, Ogata S, Clements D, Strauss JH, Baker TS, Kuhn RJ, Rossmann MG (2004) Conformational changes of the flavivirus E glycoprotein. Structure (Camb) 12:1607–1618

Zinkernagel RM, LaMarre A, Ciurea A, Hunziker L, Ochsenbein AF, McCoy KD, Fehr T, Bachmann MF, Kalinke U, Hengartner H (2001) Neutralizing antiviral antibody responses. Adv Immunol 79:1–53

Exploring the Native Human Antibody Repertoire to Create Antiviral Therapeutics

S.K. Dessain(✉), S.P. Adekar, and J.D. Berry

Abstract Native human antibodies are defined as those that arise naturally as the result of the functioning of an intact human immune system. The utility of native antibodies for the treatment of human viral diseases has been established through experience with hyperimmune human globulins. Native antibodies, as a class, differ in some respects from those obtained by recombinant library methods (phage or transgenic mouse) and possess distinct properties that may make them ideal therapeutics

S.K. Dessain
Cardeza Foundation for Hematologic Research, Kimmel Cancer Center,
Thomas Jefferson University, Curtis Building 812, 1015 Walnut St., Philadelphia, PA 19107, USA
e-mail: scott.dessain@jefferson.edu

S.K. Dessain (ed.) *Human Antibody Therapeutics for Viral Disease. Current Topics in Microbiology and Immunology 317.*
© Springer-Verlag Berlin Heidelberg 2008

for human viral diseases. Methods for cloning native human antibodies have been beset by technical problems, yet many antibodies specific for viral antigens have been cloned. In the present review, we discuss native human antibodies and ongoing improvements in cloning methods that should facilitate the creation of novel, potent antiviral therapeutics obtained from the native human antibody repertoire.

1 Introduction: The Native Human Antibody Repertoire and Viral Disease

There is a growing awareness of the utility of and need for human antibody therapeutics for viral diseases (Keller and Stiehm 2000; Oral et al. 2002; Casadevall et al. 2004; Casadevall and Pirofski 2005). Individuals who have recovered from a viral infection, or who have received a therapeutic vaccination, contain a population of antibodies that is capable of contributing to a life-long immunity from the virus. These are defined as "native antibodies," i.e., antibodies in exactly the configurations created by a functioning, intact human immune system. Native antibodies are most commonly obtained by methods that immortalize primary B cells by hybridoma formation or Epstein-Barr virus (EBV) infection. They are distinct from human or humanized antibodies derived from recombinant DNA or transgenic mouse systems, which may not accurately replicate the complete, wild-type structure of full-length antibodies produced by the human immune system in situ. The native human antibody repertoire has tremendous potential as a source for antiviral antibody therapeutics because it contains definitive immunologic solutions to human viral diseases and is likely to be the safest overall for human clinical use. Polyclonal antibody therapeutics of unselected and disease-specific native immunoglobulins are effective in some clinical situations. These intravenous immunoglobulins (IVIG) are the starting point for exploring the potential value of the native human antibody immunome, but they do not address the vast spectrum of viral diseases for which antibodies have potential therapeutic efficacy. Over the past 25 years, ongoing efforts to improve methods for cloning native human antibodies that can capture and amplify the antiviral capabilities of IVIG have progressed, demonstrating the value of this approach and justifying further exploration. In this review we will consider the history and therapeutic potential of cloned native human antibodies specific for viral illnesses.

2 Intravenous Immunoglobulins for Human Viral Disease

IVIG is the purified population of native human IgG antibodies obtained from blood plasma. The only FDA-approved antiviral use for IVIG is to treat infection by parvovirus B19 (PV B19) in patients who are immunocompromised or have aplastic anemia (Table 1). PV B19 is a small DNA virus that in normal children causes fifth disease, characterized by fever, malaise, and a typical bilateral cheek/ facial rash (Broliden et al. 2006). In most people, parvovirus infection is self-limited. However, parvovirus can cause complications in susceptible individuals resulting from its ability to infect erythroid progenitor cells. Patients who have undergone allogeneic hematopoietic stem cell transplants or who have sickle cell anemia can have an acute aplastic crisis as a result of parvovirus infection (Broliden et al. 2006; Eid et al. 2006). Fetuses carried by pregnant women infected with the virus can develop severe anemia, complicated by an infectious myocarditis, that can induce hydrops fetalis and fetal loss (Ergaz and Ornoy 2006). For these clinical situations, IVIG is often used (Moudgil et al. 1997; Geetha et al. 2000; Broliden et al. 2006; Eid et al. 2006). Human antibodies have been cloned that bind the major or minor capsid proteins and can neutralize parvovirus in vitro (Arakelov et al. 1993; Gigler et al. 1999). These may be candidates for cloned parvovirus antibody therapeutics.

Hepatitis A is a picornavirus, containing a positive-strand RNA genome, which causes an acute, self-limited hepatitis (Fiore et al. 2006). Prior to the creation of the recombinant hepatitis A vaccine, IVIG was routinely given for hepatitis A prophylaxis, although it was not FDA-approved for this purpose (CDC 1985; Fiore et al. 2006; Table 1). IVIG is still indicated for suspected, nonimmune contacts and for people who are intolerant of the vaccine. Native antibodies have been cloned that are specific for the capsid protein (Cerino et al. 1993) or the core antigen (Siemoneit et al. 1994), and one antibody has been cloned that is capable of neutralizing hepatitis A in vitro (Lewis et al. 1993).

The potential utility of IVIG for treatment of viruses other than PV B19 and hepatitis A is broad and reflects the collective antibody immunome of the population

Table 1 Common indications for the use of intravenous immunoglobulins (IVIG) or disease-specific, hyperimmune immunoglobulins (Hyper-IG). References are cited in the text

Virus	Globulins	Patients	Purpose
Parvovirus B19	IVIG	HSCT patients	Treatment
		Sickle cell disease patients	Treatment
Hepatitis A	IVIG	Anyone at risk	Prophylaxis
Hepatitis B	Hyper-IG	Possibly exposed to HBV	PEP
		Neonates	Inhibit vertical transmission
		HBV-infected liver transplant	Prophylaxis
CMV	Hyper-IG	Solid organ transplant	Treatment
VZV	Hyper-IG	Immunocompromised	Treatment of severe cases
		Neonates	Inhibit vertical transmission
VV	Hyper-IG	Immunocompromised	Treatment
Rabies	Hyper-IG	Anyone at risk	PEP
RSV	Hyper-IG	Premature infants/BPD	Prophylaxis
		HSCT patients	Treatment

BPD, bronchopulmonary dysplasia; CMV, cytomegalovirus; HBV, hepatitis B virus; HSCT, patients who have undergone allogeneic hematopoietic stem cell transplantation; PEP, post-exposure prophylaxis; VV, vaccinia virus; VZV, varicella zoster virus

from which it is derived. A study of five different IVIG preparations revealed antiviral antibodies specific for types 1, 2, 6, and 7 herpesviruses (HSV), varicella zoster (VZV), EBV, measles, mumps, rubella, and parvovirus B19 (Krause et al. 2002). Differences in the IVIGs were also noted, in that two had high levels of antibodies specific for adenovirus and two had high levels of St. Louis encephalitis virus antibodies. A panel of eight IVIG preparations was recently examined for the presence of vaccinia virus (VV) neutralizing ability (Goldsmith et al. 2004). All were found to contain significant in vitro and in vivo VV neutralizing activity, at 3%–9% of the measured titer of standard VV hyperimmune globulins (VIG), even though widespread VV vaccination has not been practiced for nearly the past 30 years. IVIG may also have clinical activity against West Nile virus (WNV). A WNV patient in Israel recovered from the infection after treatment with IVIG. It is notable that WNV is endemic in Israel and the Israeli IVIG had a high titer of WNV antibodies. In contrast, IVIG from the United States was not found to contain WNV antibodies. IVIG has also been considered to be of potential benefit for cytomegalovirus (CMV) infection in renal and bone marrow transplants (Sechet et al. 2002; Sokos et al. 2002).

3   Hyperimmune Globulins and Native Human Antibodies for Viral Diseases

The complications of some viral infections are preferably treated with hyperimmune globulins, which are polyvalent IVIGs obtained from subjects with high-titer antibody responses to specific antigens (Table 1). Some of the viruses treated with this category of therapeutics are hepatitis B (HBV), CMV, VZV, VV, rabies, and respiratory syncytial virus (RSV). The first three of these viruses share the capability of reactivation in patients who become immunocompromised. In these settings, hyperimmune globulins are used to ameliorate the immunodeficiency and bring the reactivated viruses under control. VV causes an acute infection that can have severe manifestations in immunocompromised hosts and can be mitigated by VV-specific hyperimmune globulins. Rabies induces an acute infection that is invariably fatal, unless treated, even in immunocompetent individuals. Rabies immune globulins (RIG) provide initial control of the virus while a concomitantly administered vaccine induces permanent immunity. RSV can cause fatal bronchiolitis in premature infants and recipients of allogeneic hematopoietic stem cell transplants. In these patient populations, RSV immunoglobulins (RSV-IG) are effective for RSV prophylaxis and treatment. The different roles for hyperimmune globulins are empirically defined and reflect the unique clinical features of each virus and the infected hosts. For most of these viruses, native human antibodies have been cloned that may possess some of the functions provided by the hyperimmune globulins. Future optimized hyperimmune globulins would consist of these or similar antibodies in completely defined monoclonal or oligoclonal antiviral therapies.

3.1 Hepatitis B Virus

HBV is a partially double-stranded DNA virus that is transmitted by direct contact with infected bodily fluids (Hollinger and Liang 2001). In normal individuals, hepatitis B will generally induce a self-limited hepatitis, but it has the capacity to establish a chronic active state that can eventually lead to cirrhosis or hepatocellular carcinoma. In the United States, hepatitis B immune globulin (HBIG) is prepared from a small number of donors hyperimmunized with the HBsAg vaccine (Terrault and Vyas 2003). The main three categories of use for HBIG are (1) post-exposure prophylaxis for nonimmune contacts, (2) inhibition of vertical transmission of the virus at birth, and (3) prevention of relapse in HBV-positive patients following orthotopic liver transplantation. Following a suspected infected needle-stick or fluid exposure, HBIG is recommended to be administered in combination with the HBsAg vaccine (CDC 1984). Maternal-fetal transmission can be inhibited by administration of HBIG and the HBsAg vaccination immediately after birth to infants of mothers positive for circulating HBe antigen and HBV DNA (Lo et al. 1985; Ip et al. 1989; Kabir et al. 2006). Further improvements in outcomes may also be achieved by passive immunization of HBe-positive mothers with HBIG prior to delivery (Xu et al. 2006; Xiao et al. 2007). The successful use of orthotopic liver transplantation to treat end-stage liver disease caused by HBV depends on the prevention of reactivation of the virus in the immunocompromised, post-transplant patient (Gish and McCashland 2006). HBIG can collaborate with nucleoside antiviral agents to limit reactivation in these patients, but there remains a risk of reactivation of disease after the discontinuation of prophylactic therapy. In this setting, the cost of long-term HBIG administration may potentially be reduced through the use of low-dose combination therapy regimens (Di Paolo et al. 2004; Ferretti et al. 2004).

Many native human IgG antibodies specific for HBV have been cloned using hybridoma and EBV-immortalization methods (Stricker et al. 1985; Colucci et al. 1986; Desgranges et al. 1987; Ichimori et al. 1987; Tiebout et al. 1987; Andris et al. 1992; Ehrlich et al. 1992; Sa'adu et al. 1992; Heijtink et al. 1995). Animal data are difficult to obtain with HBV infection, but a combination of two human antibodies administered to a chimpanzee chronically infected with HBV was able to transiently (< 7 days) reduce the levels of circulating virus (Heijtink et al. 1999). An important concern regarding the efficacy of cloned antibodies for HBV is the apparent ability of the virus to escape neutralization by polyclonal HBIG. This phenomenon has been observed in liver transplant patients who had recurrent HBV infection after liver transplantation despite HBIG therapy. In three studies, the existence of mutations in antigenic regions of HBsAg correlated with resistant or recurrent HBV infection (Carman et al. 1996; Ghany et al. 1998; Terrault et al. 1998). Furthermore, the length of time of therapy correlated with the likelihood of finding mutated HBV strains (Ghany et al. 1998; Terrault et al. 1998). The potential failure of polyclonal antibodies may suggest that the virus would be particularly adept at escaping the effects of a monoclonal or oligoclonal antibody therapeutic. A pair of antibodies with significant HBV binding had considerably less affinity for a variant HBV strain that had arisen in a patient following a year of HBIG therapy (Heijtink et al. 1995). Nonetheless, it may be possible to create an oligoclonal HBIG equivalent or superior to polyclonal HBIG by identifying two or three non-cross-resistant antibodies directed at relatively stable portions of HBsAg.

3.2 Cytomegalovirus

CMV is a double-stranded DNA virus that, in normal individuals, induces a febrile illness that resembles mononucleosis from EBV, with chills, fatigue, headache, and malaise (Gandhi and Khanna 2004). In immunocompromised patients, such as those who have undergone bone marrow or other organ transplantation or who have advanced human immunodeficiency virus (HIV) disease, CMV can cause considerable morbidity and mortality. CMV may be reactivated in a previously infected person who becomes immunosuppressed. Alternatively, a de novo CMV infection may be transmitted to a CMV-negative recipient of an organ from a CMV-positive donor. Many organs can be affected by CMV infection, including the retina, lung, liver, esophagus, or colon. CMV can also cause complications to a fetus infected in utero, including hearing loss, visual loss, and neurological complications (Fowler et al. 1992). CMV hyperimmune globulins (CMVIG) first demonstrated efficacy in the treatment of disease associated with kidney transplants (Snydman et al. 1987). Since then, CMVIG has been approved by the FDA for treatment of CMV reactivation in patients with transplants of the kidney, heart, lung, liver, and pancreas (Sawyer 2000). Evidence does not clearly support the use of IVIG or CMVIG in allogeneic bone marrow transplant patients (Zikos et al. 1998; Sokos et al. 2002). The utility of CMVIG in organ transplant settings has been lessened by the availability of potent small molecule anti-CMV drugs, such as ganciclovir, valganciclovir, foscarnet, and cidofovir, even though these drugs have significant toxicities (Biron 2006). It is possible that CMVIG may synergize with small molecule anti-CMV drugs in some clinical situations (Kocher et al. 2003; Varga et al. 2005; Ruttmann et al. 2006).

Many native human monoclonal antibodies specific for CMV have been described (Emanuel et al. 1984; Redmond et al. 1986; Foung et al. 1989; Bron et al. 1990; Kitamura et al. 1990; Drobyski et al. 1991; Gustafsson et al. 1991; Ohizumi et al. 1992; Ohlin et al. 1993; Rioux et al. 1994). Some of these were found to be capable of neutralizing CMV in vitro (Redmond et al. 1986; Foung et al. 1989; Ohizumi et al. 1992; Ohlin et al. 1993). The native human CMV antibody, MSL-109, has been tested for clinical efficacy (Drobyski et al. 1991). In a randomized controlled trial of allogeneic hematopoietic stem cell transplant patients, no benefit from the antibody was seen in terms of the time to development of CMV viremia or pp65 antigenemia (Boeckh et al. 2001). Studies of the MSL-109 antibody in AIDS patients with newly diagnosed or recurrent CMV retinitis did not show a reduction in the progression of CMV disease (CDC 1997a; Borucki et al. 2004). The explanation for these disappointing results is unclear.

The MSL-109 antibody is specific for the H glycoprotein (gp86). It is possible that an antibody specific for the B glycoprotein complex (gp58/116), a major target of CMV neutralizing antibodies, may be useful alone or in combination with an anti-H antibody (Ohlin et al. 1993). Nonetheless, the most likely explanation may be that T cell function is essential for CMV control in vivo and that neutralizing antibodies are minimally active in the absence of robust T cell activity (Boeckh et al. 2003).

3.3 Varicella Zoster Virus

VZV is a highly transmissible, double-stranded DNA poxvirus that induces a febrile illness (chickenpox), which is characterized in children by fever, malaise, and a pruritic, vesicular rash (CDC 1996). VZV can also reactivate in adulthood as a series of painful vesicular lesions in the distribution of a cutaneous dermatome. Infection of pregnant women during the first and second trimesters may induce the congenital varicella syndrome, which can result in significant fetal deformities, and VZV infection transmitted to newborns can be fatal (Tan and Koren 2006). Accordingly, VZV immunoglobulins (VZIG) are indicated for immunocompromised patients, pregnant women, and neonates at risk for VZV infection (CDC 2006; Tan and Koren 2006). Native human antibodies capable of neutralizing VZV in vitro have been cloned (Foung et al. 1985; Sugano et al. 1987, 1991).

3.4 Vaccinia Virus

One of the established uses of human hyperimmune globulins is the treatment of complications of vaccinia virus (Lane et al. 1969; Henderson et al. 1999). Vaccinia virus is a poxvirus that has been adapted for use as a human vaccine for the prevention of smallpox. Although generally safe for immunocompetent persons, disseminated and occasionally fatal infections can occur among patients with underlying immunodeficiencies, such as those with HIV infection, eczema, or atopic dermatitis (Henderson et al. 1999). Generalized vaccinia is a syndrome in which VV proliferation is systemically spread through the bloodstream (Redfield et al. 1987). Progressive VV infection is characterized by unrestrained proliferation of virus in the skin. Eczema vaccinatum is the excessive proliferation of VV in the skin lesions of eczema patients. For these conditions, VIG is indicated and can often lead to a complete resolution of symptoms (Henderson et al. 1999). VIG is also useful in immunocompetent individuals who have a complicated infection, such as may result from accidental infection of the periorbital region (Lewis et al. 2006).

Creation of cloned neutralizing antibody therapeutics for VV may be challenged by its complex life cycle. The VV virion exists in two forms that differ in their ability to be neutralized by antibodies, the intracellular mature virion (IMV) and the extracellular enveloped virion (EEV), with the IMV more susceptible to neutralization than the EEV (Law and Smith 2001; Smith et al. 2002). Optimal protection against lethal VV in murine and rhesus macaque models by a DNA vaccine required a combination of four genes directed at both the IMV and EEV (Hooper et al. 2003). A cloned native murine antibody specific for the A27L antigen (a neutralization target of the IMV) was able to protect mice prophylactically and therapeutically from a lethal VV challenge (Ramirez et al. 2002). However, no comparable native or nonnative human antibodies have been described. It will be important, however, to determine how many cloned human antibodies will be required to improve the symptoms of VV infection in immunocompromised patients.

3.5   Rabies Virus

Rabies is a virus with a single-stranded RNA genome that causes an acute and universally fatal encephalitis. The efficacy of RIG for the post-exposure prophylaxis of rabies has been reviewed elsewhere in this volume (see the chapter by T. Nagarajan et al.). Briefly, human rabies immunoglobulins (HRIG) are used in combination with rabies vaccination for a known or suspected rabies exposure, administered intravenously as well as directly into the suspected exposure site (see the chapter by T. Nagarajan et al., this volume). Native human antibodies have been cloned that are capable of neutralizing the virus in vitro and in vivo; most of these are reactive with the rabies glycoprotein (Dietzschold et al. 1990; Gebauer and Lindl 1990; Lafon et al. 1990; Ueki et al. 1990; Enssle et al. 1991; Dorfman et al. 1994; T. Nagarajan et al., this volume). A combination of two human antibodies that bind noncross-resistant epitopes on the glycoprotein has undergone preclinical in vivo testing. One of the antibodies was a native antibody and the other was cloned using the phage display method (Champion et al. 2000; Bakker et al. 2005). The antibody combination demonstrated efficacy comparable to HRIG and did not interfere with the potency of a simultaneously administered rabies vaccine (de Kruif et al. 2006).

3.6   Respiratory Syncytial Virus

RSV is a single-stranded, negative-strand virus that usually causes an upper respiratory infection (Welliver 2003). In some patient populations, RSV infection can develop into a bronchiolitis, an inflammation of the bronchioles, the smallest air passages of the lung. It is an important cause of mortality in young children and the elderly, and no vaccine for the disease currently exists (Shay et al. 2001; Thompson et al. 2003; Falsey et al. 2005). RSV exists in two main subtypes, A and B, but infection with one subtype does not even provide lifelong protection from reinfection by the same subtype (Welliver 2003). Premature infants and those affected by bronchopulmonary dysplasia (BPD) are at increased risk for hospitalization and death from RSV bronchiolitis (Aujard and Fauroux 2002). The prevalence of RSV infection in this population can be reduced by prophylactic treatment with RSV-IG (CDC 1997b). In contrast, RSV-IG did not show any efficacy in the treatment of infants already admitted to the hospital with the disease (Rodriguez et al. 1997). RSV may also cause a fatal bronchiolitis in patients undergoing allogeneic stem cell transplantation, and off-label administration of RSV-IG with the antiviral drug ribavirin may reduce mortality (DeVincenzo et al. 2000; Ghosh et al. 2000; Small et al. 2002).

No native human antibodies that neutralize RSV have been cloned. However, a humanized murine monoclonal antibody, palivizumab, is a potent substitute for RSV and the first demonstration of the utility of a monoclonal antibody as an antiviral therapeutic (Young 2002; see the chapter by H. Wu et al., this volume). Palivizumab binds an epitope on the F glycoprotein, a viral surface protein that is a major target for neutralizing antibodies and is highly conserved between type A and B viruses. In a series of high-risk infants with prematurity and/or BPD, a course of monthly prophylactic doses of palivizumab reduced the overall rate of serious infections and hospitalizations by 55% (CDC 1998). A role for palivizumab in the treatment of RSV infection in the elderly or in allogeneic hematopoietic stem cell patients has not yet been established.

3.7 Cloning Antibody Therapeutics for Viral Disease

The potency of IVIG, hyperimmune IGs, and the monoclonal antibody palivizumab demonstrate in principle that human antibody therapeutics are likely to be effective for the treatment of viral diseases. A vast, unmet medical need exists for treatments for the majority of viral diseases that occur worldwide. The development of antiviral antibody therapeutics will be challenged by the diversity of virus types, patient populations and the roles antibodies play in the neutralization of specific viruses, the ability of viruses to mutate antigenic domains, and an incomplete understanding of the specific features that endow an antibody with neutralizing ability. To counter these uncertainties it will be important to explore as diverse an antibody repertoire as possible, which can best be achieved by using a variety of different, complementary methods for human antibody cloning. The efficacy of IVIG and hyperimmune IGs suggests that an ideal starting point to clone a human antibody capable of potently neutralizing a viral pathogen may be with B cells from subjects who have developed a definitive antiviral body response, either by infection or vaccination. Native antibody libraries created from these affinity-matured B cells would be expected to contain individual antibodies that possess virus-neutralizing abilities and would be suitable for use as monoclonal or oligoclonal antibody therapeutics. The successes of native human antibody cloning methods in obtaining native human antiviral antibodies, and the potency of these antibodies, establish a rationale for further exploration of these methods. It is evident that the effectiveness of this approach will depend on the ability to create libraries that come as close as possible to comprehensively incorporating the entire diversity of the human antibody response to viral pathogens.

4 Features of Different Cloned Human Antibody Repertoires

The prevalent methods of cloning human antibodies from immune human repertoires differ in bias and in the degree to which they sample antibodies in their native configurations (e.g., with the original heavy chain:light chain pairing). B cell immortalization methods, which use hybridoma generation or EBV infection to enable primary human B cells to proliferate in vitro, theoretically take an unbiased sample of the repertoire of B cells and express each antibody with native heavy chain:light chain pairings. In addition to the antibodies described in the preceding section, these methods have been used to clone native human antibodies specific for measles, HIV, severe acute respiratory syndrome (SARS), EBV and hepatitis C. These methods have historically been challenged by poor antibody yields and unstable antibody secretion. Nonetheless, they have been the focus of ongoing optimization efforts that should improve their ability to comprehensively access the native human antibody immunome.

Recombinant DNA methods offer a well-established method for cloning human antibody repertoires. In these methods, heavy chain and light chain variable domains are amplified from B cell populations using RT-PCR, fused, and expressed as single-chain antigen-binding domains (scFv) on the surface of filamentous phages (Barbas 1993; Winter et al. 1994). Screening for specific antibodies is performed by panning for virus that binds to a plate or other solid support coated with antigen (Bradbury and Marks 2004). A related technology is yeast display, in which the scFv molecules are expressed on the surface of *Saccharomyces cerevisiae* (Boder and Wittrup 1997). Yeast display allows greater diversification of expressed antibody sequences by mutagenesis and has the advantage that yeast cells expressing human antibody can be directly screened by flow cytometry with fluorescent antigen, enabling a rapid assessment of binding kinetics. Libraries of scFv antibodies have also been efficiently expressed on the surface of *Escherichia coli* (Daugherty et al. 1999).

It is clear that recombinant DNA libraries obtained from immune individuals differ from antibodies obtained from nonimmune individuals (Amersdorfer et al. 2002). However, the process of creating these libraries can introduce bias at different steps in the process that may hinder their ability to capture the entire native antibody repertoire. The first step is an RT-PCR amplification with consensus DNA primers, which may not equally amplify each immunoglobulin gene sequence. The second is at the level of expression in phage, because *E. coli* does not express all eukaryotic peptides with the same efficiency, and human variable domain gene sequences can differ significantly from one another in their length and amino acid composition (Pavoni et al. 2006). A combination of these effects could potentially reduce the prevalence of specific antibodies in the antibody libraries or eliminate them entirely.

Evidence that this occurs comes from DNA sequence analysis of complementarity-determining (CDR) regions of heavy chain variable domains (V_H) cloned by phage display. The third CDR region (CDR3) of the V_H is the most important contributor to the antigen-binding specificity of an antibody (Xu and Davis 2000). CDR3 regions incorporated into phage display libraries tend to be short (less than

15 amino acids), whereas CDR3 regions in native human antibodies vary widely in length, with many over 20 amino acids in length (Griffiths et al. 1994; Brezinschek et al. 1995; Tian et al. 2007). Shorter CDR3 regions correlate overall with greater levels of somatic hypermutation, but longer CDR3 regions may be better capable of viral neutralization (Saphire et al. 2001; Hangartner et al. 2006; Tian et al. 2007). It is possible that shorter CDR3 regions are selected against at the level of PCR amplification or expression in *E. coli.*

Two studies have directly examined the types of immune libraries that arise from phage display and hybridoma methods. In one comparative study, antibodies cloned from mice immunized with human interleukin-5 protein using phage display and hybridoma methods were compared (Ames et al. 1995). Each method produced a structurally distinct group of antibodies, and only the antibodies cloned by the hybridoma method were able to block binding of the cytokine with its receptor. Ohlin and Borrebaeck (1996) analyzed a dataset of cloned antibody sequences specific for infectious disease antigens, the majority of which were viral, and were cloned by either the phage display or hybridoma method (Ohlin and Borrebaeck 1996). They noted substantial differences in the heavy chain and light chain gene family utilization between antibodies derived from the two different sources. They also noted a dramatic limitation of the diversity of the light chain gene repertoire. This observation may have been due to the phenomenon of light chain promiscuity, i.e., the ability of heavy chains to productively associate with a variety of light chains (Kang et al. 1991).

The ability of a phage library to recreate native heavy chain:light chain combinations was recently assessed by comparing a phage display antibody library that maintained native pairings with one made from the same cDNA that did not (Meijer et al. 2006). In the random library, the assortment of heavy chain and light chain sequences had apparently lost a majority of the original heavy chain:light chain pairings. Consistent with the principle of light chain promiscuity, the diversity of the random library was less than the nonrandom library due to an over-representation of VH chains capable of associating with many different light chain sequences. The functional importance of the antibody repertoire shift in the random library was revealed by the overall lower affinity of antibodies specific for tetanus toxoid (TT) antibodies cloned from the two libraries.

Taken together, these experiments illustrate the concept that intrinsic biases in phage display libraries may prevent some important native antibody structures from being incorporated into them. B cell immortalization methods of human antibody cloning are therefore complementary to recombinant DNA methods and thus merit further study and optimization.

5   Hybridoma Methods to Clone Native Human Antibodies

As a starting point in native human antibody cloning methods, the source of virus-immune B cells is an immune individual who has generated an antibody response that is effective in collaborating with the human immune system to cure the viral infection. Thus, there can be a presumption that antibodies with the requisite biological functions exist within the volunteer B cell donor. Donors can be subjects who have either (1) received vaccines specific for the virus, (2) survived an infection by the relevant virus, or (3) have succumbed to the viral infection but have made spleen, lymph nodes, or peripheral blood mononuclear cells (PBMCs) available *post mortem*. The use of B cells from a variety of genetically unrelated individuals can increase the diversity of the native antibody libraries to be screened.

There are many approaches to cloning human antibodies in their native configurations. For the most part, these involve methods of converting primary human B cells into a form that is viable in vitro through EBV immortalization, hybridoma formation, or a combination of these protocols. In EBV immortalization, purified B cells are infected with EBV-containing supernatant from the B95-8 marmoset cell line (Brown and Miller 1982). These methods are effective, yet they can be compromised by the low levels of antibody that are typically expressed by EBV-transformed cells (lymphoblastoid cells, LCLs) (Stein and Sigal 1983). In hybridoma methods, primary human B cells are fused to an immortal fusion partner cell line, which is adapted to in vitro culture and capable of producing high levels of antibody from immunoglobulin genes provided by the primary B cell. The primary impediments to hybridoma approaches have been low hybrid cell yields and the loss of antibody expression, which correlates with the loss of human chromosomes from the hybrid cells. Combination approaches have been taken that can overcome some of these defects by immortalizing and expanding the antigen-specific B cell population first with EBV infection, and then fusing the immortalized cells to a murine or murine/human fusion partner cell line.

5.1   Improvements in Fusion Partner Cell Lines

Most of the technology development in this area has attempted to address the problem of hybridoma instability by improving the fusion partner cell line. It had originally been considered that human cell lines would be optimal as fusion partners for primary human B cells because hybrid cells formed between murine cells and human cells were known to segregate human chromosomes (Ephrussi and Weiss 1969). The first reported human antibody cloning by a hybridoma method was an IgM antibody specific for measles virus (Croce et al. 1980). For this purpose, Croce et al. used a human myeloma cell line as a fusion partner cell and PBMCs from a patient with subacute sclerosing panencephalitis, the clinical syndrome resulting from measles virus infection of the central nervous system. Shortly thereafter appeared the first report of use of an EBV-immortalized B cell line as a fusion partner to clone antibodies specific for TT (Chiorazzi et al. 1982). An EBV-immortalized human B cell expressing an antibody to CMV was fused to a human myeloma cell line to give a hybrid with improved antibody expression (Emanuel et al. 1984). Enthusiasm for human cell lines was tempered, however, due to problems with the limited number of immortalized myeloma and other B cell lines that were available (Kozbor et al. 1986). Most of the cell lines had low fusion rates and produced slow-growing hybridomas, and many already expressed human antibody genes. Chromosomal instability was also observed to be a considerable problem (Olsson et al. 1983).

Experiments using murine myeloma cell lines as fusion partners for human B cells demonstrated a poor efficiency that likely resulted from the strong tendency of murine/human hybrid cells to rapidly segregate human chromosomes (Ephrussi and Weiss 1969; Schlom et al. 1980; Kozbor et al. 1982; Koropatnick et al. 1988). To compensate for the defects intrinsic to human and murine partner cell lines, a variety of heteromyeloma (murine and human) cell lines have been created. The general approach taken was to fuse murine myeloma cell lines with human cells, either normal PBMCs (Foung et al. 1984; Ichimori et al. 1985; Grunow et al. 1988) or malignant cells (Carroll et al. 1986; Posner et al. 1987; Faller et al. 1990; Shirahata et al. 1998).

Where examined, these fusions have generally resulted in hybrid cells with chimeric murine/human genomes that appear to be improved fusion partners for creating hybridoma cells that stably secrete human antibodies. For instance, the CB-F7 and the SPAM-8 heteromyelomas contained no distinct human chromosomes, but did contain human DNA detectable by hybridization analysis, probably in the form of murine/human chimeric chromosomes (Grunow et al. 1988; Gustafsson et al. 1991). The heteromyeloma cell lines K6H6/B5, HAB-1, HM-5, and SPC-H20 all possessed independent, metacentric chromosomes, consistent with a human origin (Foung et al. 1984; Ichimori et al. 1985; Carroll et al. 1986; Faller et al. 1990). When directly compared to the parental murine myeloma cell lines, the heterohybridoma fusion partner cell lines tended to have an improved ability to give rise to hybrid cells that stably expressed human antibodies (Foung et al. 1984; Carroll et al. 1986; Grunow et al. 1988; Faller et al. 1990). As many of the hybrid cells derived from these fusion partner cells contained substantial numbers of human chromosomes, it is likely the heteromyeloma cell lines were better able to produce hybrid cells with a reduced tendency to segregate human chromosomes (Foung et al. 1984; Carroll et al. 1986; Grunow et al. 1988; Faller et al. 1990). Using heteromyeloma fusion partner cell lines, a wide variety of native human antibodies have been cloned that were specific for important viral pathogens. These included the human T cell lymphotropic virus (HTLV-1), CMV, HBV, hepatitis C virus (HCV), HIV, and VZV (Foung et al. 1984; Carroll et al. 1986; Grunow et al. 1988; Bron et al. 1990; Faller et al. 1990; Gustafsson et al. 1991; Hadlock et al. 1997, 2000).

5.2 Methods of Preparing Human B Cells for Fusion

Along with the improvements in the fusion partner cell lines, the parameters affecting the rate of productive hybrid cell formation have been systematically analyzed. The best sources of primary human B cells are the splenic mononuclear cells, tonsils, or peripheral blood mononuclear cells from infants (Olsson et al. 1983; Grunow et al. 1988; Jessup et al. 2000; Karpas et al. 2001). The time of harvest of B cells following a vaccination is also important, with the best outcomes with TT antibodies seen with cells obtained 5–7 days following the vaccination, which corresponds to the period of time when the maximum quantity of TT-specific memory B cells is circulating in the blood (Butler et al. 1983; Lanzavecchia et al. 2006). Treatment of the primary B cells with a proliferative stimulus prior to fusion is also essential, either with pokeweed mitogen (PWM) or EBV (Butler et al. 1983; Larrick et al. 1983; Olsson et al. 1983; Cole et al. 1984; Emanuel et al. 1984). PWM is superior to phytohemagglutinin and is optimally used for 5–7 days (Olsson et al. 1983; Arinbjarnarson and Valdimarsson 2002). Costimulation of mitogen-treated cells with antigen can increase the yield of antigen-specific anti-bodies (Butler et al. 1983; Sugano et al. 1987). During the cell fusion, the ratio of B cells to immortal fusion partner cells is an important variable (Butler et al. 1983; Perkins et al. 1991).

Two groups of investigators have noted improvements in fusion efficiencies when the PBMCs are expanded prior to cell fusion using the CD40 system, an in vitro cell culture method that uses antibodies specific for CD40 and interleukin (IL)-4 to stimulate B cell proliferation and survival in vitro prior to cell fusion (Banchereau and Rousset 1991; Darveau et al. 1993; Thompson et al. 1994). Some of the benefit from expansion of the B cells in the CD40 system or by EBV-immortalization may derive from removing cytotoxic cells from the fusion that may threaten the viability of nascent heterohybridoma cells, which presumably express a variety of murine protein antigens, in the context of human MHC, that may be recognized as foreign by the human cytotoxic cell population. Consistent with this hypothesis, Borrebaeck and his colleagues demonstrated a dramatic improvement in the yields of murine/human and human/human cell fusions when they treated the input PBMCs with l-leucine methyl ester (Leu-OMe), which is toxic to lysosome-rich cytolytic cells, including natural killer (NK) cells and some T cells (Borrebaeck et al. 1987; Borrebaeck et al. 1988). A similar potential effect on cytotoxic, unfused cells was observed by Kalantarov et al. with the murine/human fusion partner cell line MFP-2S, which carried the *neo* drug-resistance marker. Inclusion of G418 in the cell culture medium post-fusion substantially reduced the variability of yields of antibody-secreting hybrid cells (Kalantarov et al. 2002).

In principle, it may be helpful to enrich cell populations for expression of specific antibodies prior to cell fusion. In a report approximately 20 years ago, Casali et al. selected B cells expressing antibodies specific for TT prior to EBV immortalization (Casali et al. 1986). More recently, in comparison to results obtained with unselected PBMCs, fusions performed with CD19-selected B cells had increased hybridoma yields (Schmidt et al. 2001).

5.3   *Electrofusion and Hybrid Cell Culture*

As an alternative to traditional chemical methods of inducing cell fusion with polyethylene glycol (PEG), electrofusion can offer dramatically improved rates of cell fusion (Pratt et al. 1987; Foung et al. 1990; Perkins et al. 1991). In electrofusion, the cells to be fused are aligned in a hypoosmolar buffer solution using an alternating current. Fusion is then induced by one or more bursts of direct current. Electrofusion has been used with a variety of fusion partner cell lines, including lymphoblastoid, heteromyeloma and murine myeloma cell lines (Pratt et al. 1987; Foung et al. 1990; Yoshinari et al. 1996). Three studies have directly compared the efficiency of electrofusion and polyethylene glycol (PEG) fusion, estimating an apparent superiority of electrofusion of 4- to 100-fold, with a maximal calculated fusion rate of approximately 1 cell per 1,000 input human B cells (Perkins et al. 1991; Krenn et al. 1995; Panova and Gustafsson 1995). To improve the viability of hybrid cells following fusion, a delay of 24 h prior to the initiation of HAT selection (hypoxanthine, aminopterin, thymidine) and the use of cell feeder layers to support hybridoma growth have also been found to be helpful (Cote et al. 1983; Perkins et al. 1991; Hoffmann et al. 1996; Shirahata et al. 1998).

6 Recent Advances in Native Human Antibody Cloning

Improvements continue to be made in the fusion partner cell lines. The Karpas 707H cell line is a near-tetraploid human myeloma cell line that has been specifically selected for improved growth rates in vitro and resistance to PEG, which is required for cell fusion but which was toxic to the original myeloma cell line (Karpas et al. 2001). Karpas 707H effectively fuses with tonsillar B cells and lymphoblastoid cells and is notable for the creation of hybridomas that secrete up to 210 μg antibody/ml culture medium. An analysis of the antibodies cloned from thymocytes fused to the Karpas 707H cell line revealed a spectrum of antibodies representing different stages in the B cell differentiation process (Vaisbourd et al. 2001).

MFP-2B is a novel heterohybridoma cell line that is actually the progeny of two cell fusions. The first was between a murine myeloma cell line and a human myeloma cell line. The second was between one of the resultant heterohybridomas and primary human lymphocytes obtained from a lymph node (Kalantarov et al. 2002). The MFP-2B has been additionally modified to express a *neo* resistance gene, enabling negative selection against cytotoxic cells following the cell fusion (Kalantarov et al. 2002). This cell line is notable for its fusion and cloning efficiency. A karyotype demonstrates no intact human chromosomes, but 40% of the chromosomes are partial human chromosomes or chimeric murine/human chromosomes. The MFP-2B cell line has also been used to clone antibodies specific for breast cancer antigens (Kirman et al. 2002).

An important alternative to hybridoma methods is a recently improved EBV-immortalization method, in which human primary CD19[+]IgG[+]B cells are stimulated with a CpG oligonucleotide prior to EBV exposure (Hartmann and Krieg 2000; Traggiai et al. 2004). The polyclonal B cell proliferation increases the rate of EBV immortalization from 1%–2% to 30%–100%. In addition, the efficiency of cloning the transformed cells was improved by including CpG oligonucleotides in the culture medium and using an irradiated mononuclear cell layer. Others have noted that lymphoblastoid cells tend to have unstable IgG expression, but the immortalized cells were found to produce 3–20 µg antibody/ml supernatant and were stable enough to enable in vitro and in vivo functional experiments (Stein and Sigal 1983; Traggiai et al. 2004). This method enabled the cloning of a panel of IgG antibodies specific for either the nucleoprotein or the spike protein of the SARS virus, some of which were able to significantly reduce proliferation of the virus in a murine disease model (Traggiai et al. 2004).

Improvements have also been made in phage display methods that may mitigate some of the factors that hinder the incorporation of native human antibody genes into typical phage display libraries. As described above, Meijer et al. employed a novel approach of in-cell, single-cell PCR with consensus oligonucleotides that produce an individual, correctly paired scFvs from each cell (Meijer et al. 2006). These scFvs were then used to create a phage display library for screening. Analysis of the paired sequences produced by this method demonstrated consistent pairing of the same heavy and light chains, indicating preservation of the native paired antibody conformations.

7   The Use of Ectopic Gene Expression to Improve Hybridoma Stability

Little is understood about the causes of the intrinsic instability of hybridomas formed with primary human B cells or of the segregation of human chromosomes by murine/human hybrid cells (Ephrussi and Weiss 1969; Cieplinski et al. 1983; Harris et al. 1990). To begin to address these questions empirically, we and others have considered that empiric modification of fusion partner cells by ectopic gene expression may provide insight into the nature of hybridoma cells while potentially improving their utility.

The first experiments along these lines were based on the observation that addition of IL-6 to the culture medium of murine/murine cell fusions could increase the proportion of hybrid cells expressing murine antibody (Bazin and Lemieux 1989). Interleukin-6 is essential for myeloma cell growth, possessing proliferative and antiapoptotic functions, in addition to the ability to directly stimulate antibody gene expression (Hirano 1998). Addition of IL-6 to hybridoma culture medium improved the cloning efficiency and antibody secretion of established hybridomas (Zhu et al. 1993). SP2/0 cells ectopically expressing high level mIL-6 (SP2/mIL-6) were found to give improved yields of hybridomas secreting both antigen-specific and nonspecific antibodies, compared with untransfected, parental SP2/0 cells (Harris et al. 1992).

A similar experiment was performed with the goal of improving the stability of murine/human cell fusions (Zhu et al. 1999). Interleukin-11 (IL-11), which shares many functions with IL-6, was ectopically expressed in a murine fusion partner cell line. Expression of IL-11 improved the yields of hybridomas following selection, and this effect was noted with both mitogen-stimulated and EBV-transformed

B cells. IL-11 expression also increased the quantity of antibody produced by hybrids derived from stable LCLs. However, no data were given on the long-term stability of the hybridomas. It is likely that they were still prone to segregation of human chromosomes and the associated loss of antibody expression. We performed similar experiments, comparing the ability of the SP2/mIL-6 and SP2/0 fusion partner cell lines to form stable hybrids with human splenic B cells. Expression of mIL-6 was not able to overcome the instability resulting from the segregation of human chromosomes (Dessain et al. 2004; K. Rybinski, S. Adekar, B. Barnoski, S. Dessain, unpublished data).

We originally considered that ectopic expression of human telomerase (hTERT) may improve fusions between human B cells and human immortal fusion partner cell lines. Human/human hybridomas are affected by poor proliferation rates and chromosome loss, both phenotypes having been associated in other cell culture systems with telomere dysfunction (Olsson et al. 1983; Counter et al. 1992; Bailey and Murnane 2006). Prior to the discovery of hTERT, experiments had shown that mortal human T cells impose a dominant senescence program when fused to immortal human cells (Pereira-Smith et al. 1990). Later, microcell fusion experiments revealed that the introduction of an intact copy of human chromosome 3 into an immortal, hTERT-expressing cell line repressed hTERT activity and caused cellular senescence (Oshimura and Barrett 1997). Together, these results suggested that human/human hybrid cells, formed between immortal fusion partner cell lines and primary human B cells, suffered from hTERT deficiency. Unfortunately, initial experiments with human fusion partner cell lines suggested that their deficiencies were multifactorial and could not be overcome solely by ectopic hTERT expression (S. Dessain, R. Goldsby, R. Weinberg, data not shown).

Because murine fusion partner cell lines are much better at forming hybrids than most human fusion partners, we performed similar experiments with the SP2/0 cell line (Shulman et al. 1978). The SP2/0 cell line is a very poor fusion partner for primary human B cells, so it served as a useful starting point to assess the affect of ectopic gene expression (Jessup et al. 2000). In murine/human hybrid cells, hTERT could potentially contribute to hybrid cell stability by a species-specific stabilization of human telomeres. In addition, hTERT has been shown to have many other functions that may be beneficial to hybrid cells, including an incompletely characterized tumor-promoting function that may be related to its antiapoptotic and growth factor-stimulatory activities (Holt et al. 1999; Stewart et al. 2002; Kanzaki et al. 2003; Smith et al. 2003). We found that most heterohybridoma cells formed between the SP2/0 cell line and primary human B cells expressed murine TERT (mTERT), but not hTERT. We introduced hTERT into SP2/0 cells, observing a modest increase in the numbers of cells expressing hTERT, but without useful, long-term maintenance of human antibody expression (S. Dessain, R. Goldsby, R. Weinberg, data not shown). In contrast, the ectopically expressed combination of hTERT and mIL-6 readily enabled the creation of stable hybrid cells secreting human antibodies. Notably, the hybridomas that resulted from these fusions contained considerable numbers of intact human chromosomes, even after 3 months of continuous culture in vitro (Dessain et al. 2004). An example of this is shown in Fig. 1, the human

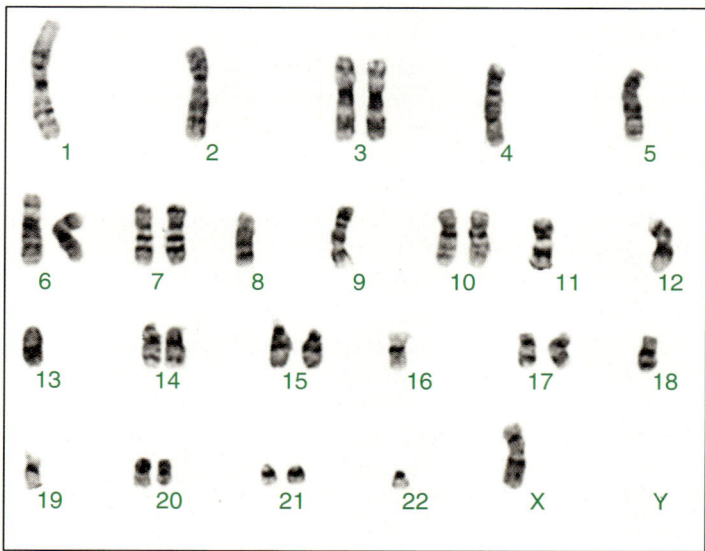

Fig. 1 Human chromosomes in a human/murine heterohybridoma. Shown are G-banded human chromosomes in a hybridoma that secretes a human antibody. Murine chromosomes are not shown. Stable antibody expression results from the ability of the hybrid cells to maintain intact human chromosomes

chromosome karyotype of a human/murine hybridoma cell that secretes a nonneutralizing IgM antibody specific for the vaccinia virus A27L antigen. Virtually a full diploid human genome is present, in addition to over 100 murine chromosomes, 10 weeks following the creation of the hybrid cell.

Some of the apparent cooperative benefit of the hTERT and mIL-6 genes may result from a mechanism whereby mIL-6 expression may promote the maintenance of human chromosomes by nascent hybridoma cells. Hybrid cells created by SP2/0 cells and human primary B cells do not proliferate in culture because the human chromosome 21 represses mIL-6 expression by the hybrid cells (Ebeling et al. 1998). Therefore, proliferation of the hybrid cells in vivo may only be possible after this chromosome is lost, indirectly selecting for cells that rapidly segregate human chromosomes. The specific mechanisms whereby mIL-6 and hTERT collaborate in chromosome maintenance are under investigation.

8   High-Throughput Screening Technologies

Following their establishment through cell fusion and drug selection, hybridomas need to be screened for specific antibody expression. In an optimal approach to thoroughly explore the native antibody immunome, each hybridoma would be assayed individually for the binding specific of its particular antibody. Because of practical considerations, hybridomas are generally assayed in pools of dozens or more clones, but advanced screening technologies may significantly increase the yields of specific antiviral antibodies that could be obtained. New antibody screening methods differ in how they achieve the core objective of associating individual cells with the antibodies they produce. In the selected lymphocyte antibody method (SLAM), primary B cells are cultured in the presence of complement and sheep red blood cells (SRBC) conjugated to the antigen of interest (Babcook et al. 1996). B cells expressing antibodies specific for the SRBC-conjugated antigen can be identified because their secreted antibodies cause localized hemolytic reactions. Although it was originally conceived that this method would be used with primary B cells from which immunoglobulin genes would be directly cloned by RT-PCR, this method may be useful for screening hybridomas or EBV-transformed cells.

Three methods combine the isolation of individual hybridoma cells with fluorescent assays for antigen binding. The first distributes individual hybridomas into tiny wells (0.1–1 nl volume) created on glass slides using a microengraving technique (Love et al. 2006). The secreted antibodies are captured for analysis by sandwiching the arrayed hybridoma supernatants with a capture slide that is coated with secondary antibody or antigen. The bound complexes are then detected with fluorescently labeled antigens or secondary antibodies, respectively. Multiple capture slides can be used with a single hybridoma microarray, enabling cells to be screened for a variety of antigen-binding specificities. An alternate method immobilizes hybridomas on a filter through which secreted antibodies diffuse and then bind to a plate coated with a secondary antibody (www.trellisbio.com; Potera 2005). The plate is then probed with a panel of fluorescent probes that can be used in a combinatorial fashion to allow simultaneous screening for many different antigens. Computerized microscopy is used to analyze the binding reactions. Hybridomas can also be enveloped in an agarose matrix that captures the antibodies secreted by the hybridomas. For this purpose, secondary antibodies are attached to the agarose through a biotin-avidin bridge (Gray et al. 1995). The secreted antibodies are thus stably associated with the cells that produce them. The porous agarose matrix enables the hybridomas to be screened for binding to fluorescently labeled antigens. The matrix also offers structural stabilization for the hybridomas so that they can be analyzed and sorted by FACS.

In one of our laboratories (J.B.) we have begun experiments with the FMAT 8200 Cellular Detection System. This system uses antigen-coated beads, which are mixed with hybridoma supernatants and fluorescently labeled secondary antibodies in 96-well or 386-well formats. The secondary antibodies detect specific antibody bound to the beads, thereby concentrating the fluorescence into punctate signals that are detected by a mechanized plate reader. The advantage of this method is that it enables high-throughput screening of hybridoma supernatants, but it does not provide a means of isolating single cells prior to screening. Finally, the marriage of such cell screening technologies with automated cell manipulators (ClonePix, www.genetix.com) will accelerate the process of mining the native human antibody immunome to obtain antibodies for use in the treatment of viral diseases.

9 Summary and Future Prospects

The pressing demand for antibodies for use as antiviral therapeutics mandates a broad-based effort that utilizes all of the available antibody cloning technologies in parallel. Cloning methods that directly immortalize B cells through hybridoma creation or EBV infection can complement recombinant DNA and transgenic mouse methods of antibody cloning because they have an unbiased access to antibody repertoires in their native configurations. In addition, these methods simplify the exploration of the antibody repertoires of genetically diverse individuals. Over the past 27 years, successive technical advances have improved the methods for cloning native human antibodies such that they now may be able to contribute meaningfully to ongoing efforts with phage display and transgenic mouse methods. The simplicity of these methods should facilitate their application by laboratories with a diversity of research interests, as well as provide a rationale for creating core facilities that provide high-throughput screening services to academic and other researchers.

Acknowledgements We are grateful to Mary Guttieri and Jay Hooper of USAMRIID, Frederick, MD, for reagents used to clone the VV A27L hybridoma described in Fig. 1. We also thank Maria L. Skorski and Barry L. Barnoski for the karyotype shown in Fig. 1. For guidance and discussion we thank Hilary Koprowski, Richard Goldsby, Lance Simpson, Gordon Freeman, T. Nagarajan, Ike Eisenlohr, and Tara Robinson. This work was supported by NIH grants R01-HL081503 and R01-AI065967 (S.K.D.) and the CBRN Research Technology Initiative (J.D.B.).

References

Amersdorfer P, Wong C, Smith T, Chen S, Deshpande S, Sheridan R, Marks JD (2002) Genetic and immunological comparison of anti-botulinum type A antibodies from immune and non-immune human phage libraries. Vaccine 20:1640–1648

Ames RS, Tornetta MA, McMillan LJ, Kaiser KF, Holmes SD, Appelbaum E, Cusimano DM, Theisen TW, Gross MS, Jones CS, et al (1995) Neutralizing murine monoclonal antibodies to human IL-5 isolated from hybridomas and a filamentous phage Fab display library. J Immunol 154:6355–6364

Andris JS, Ehrlich PH, Ostberg L, Capra JD (1992) Probing the human antibody repertoire to exogenous antigens. Characterization of the H and L chain V region gene segments from anti-hepatitis B virus antibodies. J Immunol 149:4053–4059

Arakelov S, Gorny MK, Williams C, Riggin CH, Brady F, Collett MS, Zolla-Pazner S (1993) Generation of neutralizing anti-B19 parvovirus human monoclonal antibodies from patients infected with human immunodeficiency virus. J Infect Dis 168:580–585

Arinbjarnarson S, Valdimarsson H (2002) Generation of heterohybridomas secreting human immunoglobulins; pokeweed mitogen prestimulation is highly effective but phytohemagglutinin drives most B cells into apoptosis. J Immunol Methods 259:139–148

Aujard Y, Fauroux B (2002) Risk factors for severe respiratory syncytial virus infection in infants. Respir Med 96 [Suppl B]:S9–S14

Babcook JS, Leslie KB, Olsen OA, Salmon RA, Schrader JW (1996) A novel strategy for generating monoclonal antibodies from single, isolated lymphocytes producing antibodies of defined specificities. Proc Natl Acad Sci U S A 93:7843–7848

Bailey SM, Murnane JP (2006) Telomeres, chromosome instability and cancer. Nucleic Acids Res 34:2408–2417

Bakker AB, Marissen WE, Kramer RA, Rice AB, Weldon WC, Niezgoda M, Hanlon CA, Thijsse S, Backus HH, de Kruif J, Dietzschold B, Rupprecht CE, Goudsmit J (2005) Novel human monoclonal antibody combination effectively neutralizing natural rabies virus variants and individual in vitro escape mutants. J Virol 79:9062–9068

Banchereau J, Rousset F (1991) Growing human B lymphocytes in the CD40 system. Nature 353:678–679

Barbas CF 3rd (1993) Recent advances in phage display. Curr Opin Biotechnol 4:526–530

Bazin R, Lemieux R (1989) Increased proportion of B cell hybridomas secreting monoclonal antibodies of desired specificity in cultures containing macrophage-derived hybridoma growth factor (IL-6). J Immunol Methods 116:245–249

Biron KK (2006) Antiviral drugs for cytomegalovirus diseases. Antiviral Res 71:154–163

Boder ET, Wittrup KD (1997) Yeast surface display for screening combinatorial polypeptide libraries. Nat Biotechnol 15:553–557

Boeckh M, Bowden RA, Storer B, Chao NJ, Spielberger R, Tierney DK, Gallez-Hawkins G, Cunningham T, Blume KG, Levitt D, Zaia JA (2001) Randomized, placebo-controlled, double-blind study of a cytomegalovirus-specific monoclonal antibody (MSL-109) for prevention of cytomegalovirus infection after allogeneic hematopoietic stem cell transplantation. Biol Blood Marrow Transplant 7:343–351

Boeckh M, Leisenring W, Riddell SR, Bowden RA, Huang ML, Myerson D, Stevens-Ayers T, Flowers ME, Cunningham T, Corey L (2003) Late cytomegalovirus disease and mortality in recipients of allogeneic hematopoietic stem cell transplants: importance of viral load and T-cell immunity. Blood 101:407–414

Borrebaeck CA, Danielsson L, Moller SA (1987) Human monoclonal antibodies produced from L-leucine methyl ester-treated and in vitro immunized peripheral blood lymphocytes. Biochem Biophys Res Commun 148:941–946

Borrebaeck CA, Danielsson L, Moller SA (1988) Human monoclonal antibodies produced by primary in vitro immunization of peripheral blood lymphocytes. Proc Natl Acad Sci USA 85:3995–3999

Borucki MJ, Spritzler J, Asmuth DM, Gnann J, Hirsch MS, Nokta M, Aweeka F, Nadler PI, Sattler F, Alston B, Nevin TT, Owens S, Waterman K, Hubbard L, Caliendo A, Pollard RB (2004) A phase II, double-masked, randomized, placebo-controlled evaluation of a human monoclonal anti-Cytomegalovirus antibody (MSL-109) in combination with standard therapy versus standard therapy alone in the treatment of AIDS patients with cytomegalovirus retinitis. Antiviral Res 64:103–111

Bradbury AR, Marks JD (2004) Antibodies from phage antibody libraries. J Immunol Methods 290:29–49

Brezinschek HP, Brezinschek RI, Lipsky PE (1995) Analysis of the heavy chain repertoire of human peripheral B cells using single-cell polymerase chain reaction. J Immunol 155:190–202

Broliden K, Tolfvenstam T, Norbeck O (2006) Clinical aspects of parvovirus B19 infection. J Intern Med 260:285–304

Bron D, Delforge A, Lagneaux L, De Martynoff G, Bosmans E, Van der Auwera P, Snoeck R, Burny A, Stryckmans P (1990) Production of human monoclonal IgG antibodies reacting with cytomegalovirus (CMV). J Immunol Methods 130:209–216

Brown NA, Miller G (1982) Immunoglobulin expression by human B lymphocytes clonally transformed by Epstein Barr virus. J Immunol 128:24–29

Butler JL, Lane HC, Fauci AS (1983) Delineation of optimal conditions for producing mouse-human heterohybridomas from human peripheral blood B cells of immunized subjects. J Immunol 130:165–168

Carman WF, Trautwein C, van Deursen FJ, Colman K, Dornan E, McIntyre G, Waters J, Kliem V, Muller R, Thomas HC, Manns MP (1996) Hepatitis B virus envelope variation after transplantation with and without hepatitis B immune globulin prophylaxis. Hepatology 24:489–493

Carroll WL, Thielemans K, Dilley J, Levy R (1986) Mouse x human heterohybridomas as fusion partners with human B cell tumors. J Immunol Methods 89:61–72

Casadevall A, Pirofski LA (2005) The potential of antibody-mediated immunity in the defence against biological weapons. Expert Opin Biol Ther 5:1359–1372

Casadevall A, Dadachova E, Pirofski LA (2004) Passive antibody therapy for infectious diseases. Nat Rev Microbiol 2:695–703

Casali P, Inghirami G, Nakamura M, Davies TF, Notkins AL (1986) Human monoclonals from antigen-specific selection of B lymphocytes and transformation by EBV. Science 234:476–479

CDC (1984) Postexposure prophylaxis of hepatitis B. Morb Mortal Wkly Rep 33:285–290

CDC (1985) Recommendations for protection against viral hepatitis. Morb Mortal Wkly Rep 34:313–324, 329–335

CDC (1996) Prevention of varicella: recommendations of the Advisory Committee on Immunization Practices (ACIP). Centers for Disease Control and Prevention. MMWR Recomm Rep 45:1–36

CDC (1997a) MSL-109 adjuvant therapy for cytomegalovirus retinitis in patients with acquired immunodeficiency syndrome: the Monoclonal Antibody Cytomegalovirus Retinitis Trial. The Studies of Ocular Complications of AIDS Research Group. AIDS Clinical Trials Group. Arch Ophthalmol 115:1528–1536

CDC (1997b) Reduction of respiratory syncytial virus hospitalization among premature infants and infants with bronchopulmonary dysplasia using respiratory syncytial virus immune globulin prophylaxis. The PREVENT Study Group. Pediatrics 99:93–99

CDC (1998) Palivizumab, a humanized respiratory syncytial virus monoclonal antibody, reduces hospitalization from respiratory syncytial virus infection in high-risk infants. The IMpact-RSV Study Group. Pediatrics 102:531–537

CDC (2006) A new product (VariZIG) for postexposure prophylaxis of varicella available under an investigational new drug application expanded access protocol. MMWR Morb Mortal Wkly Rep 55:209–210

Cerino A, Boender P, La Monica N, Rosa C, Habets W, Mondelli MU (1993) A human monoclonal antibody specific for the N terminus of the hepatitis C virus nucleocapsid protein. J Immunol 151:7005–7015

Champion JM, Kean RB, Rupprecht CE, Notkins AL, Koprowski H, Dietzschold B, Hooper DC (2000) The development of monoclonal human rabies virus-neutralizing antibodies as a substitute for pooled human immune globulin in the prophylactic treatment of rabies virus exposure. J Immunol Methods 235:81–90

Chiorazzi N, Wasserman RL, Kunkel HG (1982) Use of Epstein-Barr virus-transformed B cell lines for the generation of immunoglobulin-producing human B cell hybridomas. J Exp Med 156:930–935

Cieplinski W, Reardon P, Testa MA (1983) Non-random human chromosome distribution in human-mouse myeloma somatic cell hybrids. Cytogenet Cell Genet 35:93–99

Cole SP, Campling BG, Louwman IH, Kozbor D, Roder JC (1984) A strategy for the production of human monoclonal antibodies reactive with lung tumor cell lines. Cancer Res 44:2750–2753

Colucci G, Kohtz DS, Waksal SD (1986) Preparation and characterization of human monoclonal antibodies directed against the hepatitis B virus surface antigen. Liver 6:145–152

Cote RJ, Morrissey DM, Houghton AN, Beattie EJ Jr, Oettgen HF, Old LJ (1983) Generation of human monoclonal antibodies reactive with cellular antigens. Proc Natl Acad Sci USA 80:2026–2030

Counter CM, Avilion AA, LeFeuvre CE, Stewart NG, Greider CW, Harley CB, Bacchetti S (1992) Telomere shortening associated with chromosome instability is arrested in immortal cells which express telomerase activity. EMBO J 11:1921–1929

Croce CM, Linnenbach A, Hall W, Steplewski Z, Koprowski H (1980) Production of human hybridomas secreting antibodies to measles virus. Nature 288:488–489

Darveau A, Chevrier MC, Neron S, Delage R, Lemieux R (1993) Efficient preparation of human monoclonal antibody-secreting heterohybridomas using peripheral B lymphocytes cultured in the CD40 system. J Immunol Methods 159:139–143

Daugherty PS, Olsen MJ, Iverson BL, Georgiou G (1999) Development of an optimized expression system for the screening of antibody libraries displayed on the Escherichia coli surface. Protein Eng 12:613–621

de Kruif J, Bakker AB, Marissen WE, Kramer RA, Throsby M, Rupprecht CE, Goudsmit J (2006) A human monoclonal antibody cocktail as a novel component of rabies postexposure prophylaxis. Annu Rev Med 58:359–368

Desgranges C, Paire J, Pichoud C, Souche S, Frommel D, Trepo C (1987) High affinity human monoclonal antibodies directed against hepatitis B surface antigen. J Virol Methods 16:281–292

Dessain SK, Adekar SP, Stevens JB, Carpenter KA, Skorski ML, Barnoski BL, Goldsby RA, Weinberg RA (2004) High efficiency creation of human monoclonal antibody-producing hybridomas. J Immunol Methods 291:109–122

DeVincenzo JP, Hirsch RL, Fuentes RJ, Top FH Jr (2000) Respiratory syncytial virus immune globulin treatment of lower respiratory tract infection in pediatric patients undergoing bone marrow transplantation—a compassionate use experience. Bone Marrow Transplant 25:161–165

Di Paolo D, Tisone G, Piccolo P, Lenci I, Zazza S, Angelico M (2004) Low-dose hepatitis B immunoglobulin given "on demand" in combination with lamivudine: a highly cost-effective approach to prevent recurrent hepatitis B virus infection in the long-term follow-up after liver transplantation. Transplantation 77:1203–1208

Dietzschold B, Gore M, Casali P, Ueki Y, Rupprecht CE, Notkins AL, Koprowski H (1990) Biological characterization of human monoclonal antibodies to rabies virus. J Virol 64:3087–3090

Dorfman N, Dietzschold B, Kajiyama W, Fu ZF, Koprowski H, Notkins AL (1994) Development of human monoclonal antibodies to rabies. Hybridoma 13:397–402

Drobyski WR, Gottlieb M, Carrigan D, Ostberg L, Grebenau M, Schran H, Magid P, Ehrlich P, Nadler PI, Ash RC (1991) Phase I study of safety and pharmacokinetics of a human anticytomegalovirus monoclonal antibody in allogeneic bone marrow transplant recipients. Transplantation 51:1190–1196

Ebeling SB, Bos HM, Slater R, Overkamp WJ, Cuthbert AP, Newbold RF, Zdzienicka MZ, Aarden LA (1998) Human chromosome 21 determines growth factor dependence in human/mouse B-cell hybridomas. Cancer Res 58:2863–2868

Ehrlich PH, Moustafa ZA, Justice JC, Harfeldt KE, Kelley RL, Ostberg L (1992) Characterization of human monoclonal antibodies directed against hepatitis B surface antigen. Hum Antibodies Hybridomas 3:2–7

Eid AJ, Brown RA, Patel R, Razonable RR (2006) Parvovirus B19 infection after transplantation: a review of 98 cases. Clin Infect Dis 43:40–48

Emanuel D, Gold J, Colacino J, Lopez C, Hammerling U (1984) A human monoclonal antibody to cytomegalovirus (CMV). J Immunol 133:2202–2205

Enssle K, Kurrle R, Kohler R, Muller H, Kanzy EJ, Hilfenhaus J, Seiler FR (1991) A rabies-specific human monoclonal antibody that protects mice against lethal rabies. Hybridoma 10:547–556

Ephrussi B, Weiss MC (1969) Hybrid somatic cells. Sci Am 220:26–35

Ergaz Z, Ornoy A (2006) Parvovirus B19 in pregnancy. Reprod Toxicol 21:421–435

Faller G, Vollmers HP, Weiglein I, Marx A, Zink C, Pfaff M, Muller-Hermelink HK (1990) HAB-1, a new heteromyeloma for continuous production of human monoclonal antibodies. Br J Cancer 62:595–598

Falsey AR, Hennessey PA, Formica MA, Cox C, Walsh EE (2005) Respiratory syncytial virus infection in elderly and high-risk adults. N Engl J Med 352:1749–1759

Ferretti G, Merli M, Ginanni Corradini S, Callejon V, Tanzilli P, Masini A, Ferretti S, Iappelli M, Rossi M, Rivanera D, Lilli D, Mancini C, Attili A, Berloco P (2004) Low-dose intramuscular hepatitis B immune globulin and lamivudine for long-term prophylaxis of hepatitis B recurrence after liver transplantation. Transplant Proc 36:535–538

Fiore AE, Wasley A, Bell BP (2006) Prevention of hepatitis A through active or passive immunization: recommendations of the Advisory Committee on Immunization Practices (ACIP). MMWR Recomm Rep 55:1–23

Foung S, Perkins S, Kafadar K, Gessner P, Zimmermann U (1990) Development of microfusion techniques to generate human hybridomas. J Immunol Methods 134:35–42

Foung SK, Perkins S, Raubitschek A, Larrick J, Lizak G, Fishwild D, Engleman EG, Grumet FC (1984) Rescue of human monoclonal antibody production from an EBV-transformed B cell line by fusion to a human-mouse hybridoma. J Immunol Methods 70:83–90

Foung SK, Perkins S, Koropchak C, Fishwild DM, Wittek AE, Engleman EG, Grumet FC, Arvin AM (1985) Human monoclonal antibodies neutralizing varicella-zoster virus. J Infect Dis 152:280–285

Foung SK, Perkins S, Bradshaw P, Rowe J, Rabin LB, Reyes GR, Lennette ET (1989) Human monoclonal antibodies to human cytomegalovirus. J Infect Dis 159:436–443

Fowler KB, Stagno S, Pass RF, Britt WJ, Boll TJ, Alford CA (1992) The outcome of congenital cytomegalovirus infection in relation to maternal antibody status. N Engl J Med 326:663–667

Gandhi MK, Khanna R (2004) Human cytomegalovirus: clinical aspects, immune regulation, and emerging treatments. Lancet Infect Dis 4:725–738

Gebauer W, Lindl T (1990) Construction of human monoclonal antibodies against rabies NS-protein antigen by Epstein-Barr virus transformation. Arzneimittelforschung 40:718–722

Geetha D, Zachary JB, Baldado HM, Kronz JD, Kraus ES (2000) Pure red cell aplasia caused by Parvovirus B19 infection in solid organ transplant recipients: a case report and review of literature. Clin Transplant 14:586–591

Ghany MG, Ayola B, Villamil FG, Gish RG, Rojter S, Vierling JM, Lok AS (1998) Hepatitis B virus S mutants in liver transplant recipients who were reinfected despite hepatitis B immune globulin prophylaxis. Hepatology 27:213–222

Ghosh S, Champlin RE, Englund J, Giralt SA, Rolston K, Raad I, Jacobson K, Neumann J, Ippoliti C, Mallik S, Whimbey E (2000) Respiratory syncytial virus upper respiratory tract illnesses in adult blood and marrow transplant recipients: combination therapy with aerosolized ribavirin and intravenous immunoglobulin. Bone Marrow Transplant 25:751–755

Gigler A, Dorsch S, Hemauer A, Williams C, Kim S, Young NS, Zolla-Pazner S, Wolf H, Gorny MK, Modrow S (1999) Generation of neutralizing human monoclonal antibodies against parvovirus B19 proteins. J Virol 73:1974–1979

Gish RG, McCashland T (2006) Hepatitis B in liver transplant recipients. Liver Transpl 12: S54–S64

Goldsmith JC, Eller N, Mikolajczyk M, Manischewitz J, Golding H, Scott DE (2004) Intravenous immunoglobulin products contain neutralizing antibodies to vaccinia. Vox Sang 86:125–129

Gray F, Kenney JS, Dunne JF (1995) Secretion capture and report web: use of affinity derivatized agarose microdroplets for the selection of hybridoma cells. J Immunol Methods 182:155–163

Griffiths AD, Williams SC, Hartley O, Tomlinson IM, Waterhouse P, Crosby WL, Kontermann RE, Jones PT, Low NM, Allison TJ, et al (1994) Isolation of high affinity human antibodies directly from large synthetic repertoires. EMBO J 13:3245–3260

Grunow R, Jahn S, Porstmann T, Kiessig SS, Steinkellner H, Steindl F, Mattanovich D, Gurtler L, Deinhardt F, Katinger H, et al (1988) The high efficiency, human B cell immortalizing hetero-myeloma CB-F7. Production of human monoclonal antibodies to human immunodeficiency virus. J Immunol Methods 106:257–265

Gustafsson B, Jondal M, Sundqvist VA (1991) SPAM-8, a mouse-human heteromyeloma fusion partner in the production of human monoclonal antibodies. Establishment of a human mono-clonal antibody against cytomegalovirus. Hum Antibodies Hybridomas 2:26–32

Hadlock KG, Rowe J, Perkins S, Bradshaw P, Song GY, Cheng C, Yang J, Gascon R, Halmos J, Rehman SM, McGrath MS, Foung SK (1997) Neutralizing human monoclonal antibodies to conformational epitopes of human T-cell lymphotropic virus type 1 and 2 gp46. J Virol 71:5828–5840

Hadlock KG, Lanford RE, Perkins S, Rowe J, Yang Q, Levy S, Pileri P, Abrignani S, Foung SK (2000) Human monoclonal antibodies that inhibit binding of hepatitis C virus E2 protein to CD81 and recognize conserved conformational epitopes. J Virol 74:10407–10416

Hangartner L, Zinkernagel RM, Hengartner H (2006) Antiviral antibody responses: the two extremes of a wide spectrum. Nat Rev Immunol 6:231–243

Harris JF, Koropatnick J, Pearson J (1990) Spontaneous and radiation-induced genetic instability of heteromyeloma hybridoma cells. Mol Biol Med 7:485–493

Harris JF, Hawley RG, Hawley TS, Crawford-Sharpe GC (1992) Increased frequency of both total and specific monoclonal antibody producing hybridomas using a fusion partner that constitutively expresses recombinant IL-6. J Immunol Methods 148:199–207

Hartmann G, Krieg AM (2000) Mechanism and function of a newly identified CpG DNA motif in human primary B cells. J Immunol 164:944–953

Heijtink R, Paulij W, van Bergen P, van Roosmalen M, Rohm D, Eichentopf B, Muchmore E, de Man R, Osterhaus A (1999) In vivo activity of a mixture of two human monoclonal antibodies (anti-HBs) in a chronic hepatitis B virus carrier chimpanzee. J Gen Virol 80:1529–1535

Heijtink RA, Kruining J, Weber YA, de Man RA, Schalm SW (1995) Anti-hepatitis B virus activity of a mixture of two monoclonal antibodies in an "inhibition in solution" assay. Hepatology 22:1078–1083

Henderson DA, Inglesby TV, Bartlett JG, Ascher MS, Eitzen E, Jahrling PB, Hauer J, Layton M, McDade J, Osterholm MT, O'Toole T, Parker G, Perl T, Russell PK, Tonat K (1999) Smallpox as a biological weapon: medical and public health management. Working Group on Civilian Biodefense. JAMA 281:2127–2137

Hirano T (1998) Interleukin 6 and its receptor: ten years later. Int Rev Immunol 16:249–284

Hoffmann P, Jimenez-Diaz M, Weckesser J, Bessler WG (1996) Murine bone marrow-derived macrophages constitute feeder cells for human B cell hybridomas. J Immunol Methods 196:85–91

Hollinger FB, Liang TJ (2001) Hepatitis B virus. In: Knipe DM, Howley PM (eds) Fields virology. Lippincott Williams and Wilkins, Philadelphia, pp 2971–3036

Holt SE, Glinsky VV, Ivanova AB, Glinsky GV (1999) Resistance to apoptosis in human cells conferred by telomerase function and telomere stability. Mol Carcinog 25:241–248

Hooper JW, Custer DM, Thompson E (2003) Four-gene-combination DNA vaccine protects mice against a lethal vaccinia virus challenge and elicits appropriate antibody responses in nonhuman primates. Virology 306:181–195

Ichimori Y, Sasano K, Itoh H, Hitotsumachi S, Kimura Y, Kaneko K, Kida M, Tsukamoto K (1985) Establishment of hybridomas secreting human monoclonal antibodies against tetanus toxin and hepatitis B virus surface antigen. Biochem Biophys Res Commun 129:26–33

Ichimori Y, Harada K, Hitotsumachi S, Tsukamoto K (1987) Establishment of hybridoma secreting human monoclonal antibody against hepatitis B virus surface antigen. Biochem Biophys Res Commun 142:805–812

Ip HM, Lelie PN, Wong VC, Kuhns MC, Reesink HW (1989) Prevention of hepatitis B virus carrier state in infants according to maternal serum levels of HBV DNA. Lancet 1:406–410

Jessup CF, Baxendale H, Goldblatt D, Zola H (2000) Preparation of human-mouse heterohybridomas against an immunising antigen. J Immunol Methods 246:187–202

Kabir A, Alavian SM, Ahanchi N, Malekzadeh R (2006) Combined passive and active immunoprophylaxis for preventing perinatal transmission of the hepatitis B virus in infants born to HBsAg positive mothers in comparison with vaccine alone. Hepatol Res 36:265–271

Kalantarov GF, Rudchenko SA, Lobel L, Trakht I (2002) Development of a fusion partner cell line for efficient production of human monoclonal antibodies from peripheral blood lymphocytes. Hum Antibodies 11:85–96

Kang AS, Jones TM, Burton DR (1991) Antibody redesign by chain shuffling from random combinatorial immunoglobulin libraries. Proc Natl Acad Sci U S A 88:11120–11123

Kanzaki Y, Onoue F, Sakurai H, Ide T (2003) Telomerase upregulates expression levels of interleukin (IL)-1alpha, IL-1beta, IL-6, IL-8, and granulocyte-macrophage colony-stimulating factor in normal human fibroblasts. Biochem Biophys Res Commun 305:150–154

Karpas A, Dremucheva A, Czepulkowski BH (2001) A human myeloma cell line suitable for the generation of human monoclonal antibodies. Proc Natl Acad Sci U S A 98:1799–1804

Keller MA, Stiehm ER (2000) Passive immunity in prevention and treatment of infectious diseases. Clin Microbiol Rev 13:602–614

Kirman I, Kalantarov GF, Lobel LI, Hibshoosh H, Estabrook A, Canfield R, Trakht I (2002) Isolation of native human monoclonal autoantibodies to breast cancer. Hybrid Hybridomics 21:405–414

Kitamura K, Yamada K, Kuzushima K, Morishima T, Morishima Y, Yamamoto N, Nishiyama Y, Ohya K, Yamaguchi H (1990) Human monoclonal antibodies to human cytomegalovirus derived from peripheral blood lymphocytes of healthy adults. J Med Virol 32:60–66

Kocher AA, Bonaros N, Dunkler D, Ehrlich M, Schlechta B, Zweytick B, Grimm M, Zuckermann A, Wolner E, Laufer G (2003) Long-term results of CMV hyperimmune globulin prophylaxis in 377 heart transplant recipients. J Heart Lung Transplant 22:250–257

Koropatnick J, Pearson J, Harris JF (1988) Extensive loss of human DNA accompanies loss of antibody production in heteromyeloma hybridoma cells. Mol Biol Med 5:69–83

Kozbor D, Roder JC, Chang TH, Steplewski Z, Koprowski H (1982) Human anti-tetanus toxoid monoclonal antibody secreted by EBV-transformed human B cells fused with murine myeloma. Hybridoma 1:323–328

Kozbor D, Roder JC, Sierzega ME, Cole SP, Croce CM (1986) Comparative phenotypic analysis of available human hybridoma fusion partners. Methods Enzymol 121:120–140

Krause I, Wu R, Sherer Y, Patanik M, Peter JB, Shoenfeld Y (2002) In vitro antiviral and antibacterial activity of commercial intravenous immunoglobulin preparations—a potential role for adjuvant intravenous immunoglobulin therapy in infectious diseases. Transfus Med 12:133–139

Krenn V, von Landenberg P, Wozniak E, Kissler C, Hermelink HK, Zimmermann U, Vollmers HP (1995) Efficient immortalization of rheumatoid synovial tissue B-lymphocytes. A comparison between the techniques of electric field-induced and PEG fusion. Hum Antibodies Hybridomas 6:47–51

Lafon M, Edelman L, Bouvet JP, Lafage M, Montchatre E (1990) Human monoclonal antibodies specific for the rabies virus glycoprotein and N protein. J Gen Virol 71:1689–1696

Lane JM, Ruben FL, Neff JM, Millar JD (1969) Complications of smallpox vaccination, 1968. N Engl J Med 281:1201–1208

Lanzavecchia A, Bernasconi N, Traggiai E, Ruprecht CR, Corti D, Sallusto F (2006) Understanding and making use of human memory B cells. Immunol Rev 211:303–309

Larrick JW, Truitt KE, Raubitschek AA, Senyk G, Wang JC (1983) Characterization of human hybridomas secreting antibody to tetanus toxoid. Proc Natl Acad Sci U S A 80:6376–6380

Law M, Smith GL (2001) Antibody neutralization of the extracellular enveloped form of vaccinia virus. Virology 280:132–142

Lewis AP, Lemon SM, Barber KA, Murphy P, Parry NR, Peakman TC, Sims MJ, Worden J, Crowe JS (1993) Rescue, expression, and analysis of a neutralizing human anti-hepatitis A virus monoclonal antibody. J Immunol 151:2829–2838

Lewis FM, Chernak E, Goldman E, Li Y, Karem K, Damon IK, Henkel R, Newbern EC, Ross P, Johnson CC (2006) Ocular vaccinia infection in laboratory worker, Philadelphia, 2004. Emerg Infect Dis 12:134–137

Lo KJ, Tsai YT, Lee SD, Wu TC, Wang JY, Chen GH, Yeh CL, Chiang BN, Yeh SH, Goudeau A, et al (1985) Immunoprophylaxis of infection with hepatitis B virus in infants born to hepatitis B surface antigen-positive carrier mothers. J Infect Dis 152:817–822

Love JC, Ronan JL, Grotenbreg GM, van der Veen AG, Ploegh HL (2006) A microengraving method for rapid selection of single cells producing antigen-specific antibodies. Nat Biotechnol 24:703–707

Meijer PJ, Andersen PS, Haahr Hansen M, Steinaa L, Jensen A, Lantto J, Oleksiewicz MB, Tengbjerg K, Poulsen TR, Coljee VW, Bregenholt S, Haurum JS, Nielsen LS (2006) Isolation of human antibody repertoires with preservation of the natural heavy and light chain pairing. J Mol Biol 358:764–772

Moudgil A, Shidban H, Nast CC, Bagga A, Aswad S, Graham SL, Mendez R, Jordan SC (1997) Parvovirus B19 infection-related complications in renal transplant recipients: treatment with intravenous immunoglobulin. Transplantation 64:1847–1850

Ohizumi Y, Suzuki H, Matsumoto Y, Masuho Y, Numazaki Y (1992) Neutralizing mechanisms of two human monoclonal antibodies against human cytomegalovirus glycoprotein 130/55. J Gen Virol 73:2705–2707

Ohlin M, Borrebaeck CA (1996) Characteristics of human antibody repertoires following active immune responses in vivo. Mol Immunol 33:583–592

Ohlin M, Sundqvist VA, Mach M, Wahren B, Borrebaeck CA (1993) Fine specificity of the human immune response to the major neutralization epitopes expressed on cytomegalovirus gp58/116 (gB), as determined with human monoclonal antibodies. J Virol 67:703–710

Olsson L, Kronstrom H, Cambon-De Mouzon A, Honsik C, Brodin T, Jakobsen B (1983) Antibody producing human-human hybridomas. I. Technical aspects. J Immunol Methods 61:17–32

Oral HB, Ozakin C, Akdis CA (2002) Back to the future: antibody-based strategies for the treatment of infectious diseases. Mol Biotechnol 21:225–239

Oshimura M, Barrett JC (1997) Multiple pathways to cellular senescence: role of telomerase repressors. Eur J Cancer 33:710–715

Panova I, Gustafsson B (1995) Increased human hybridoma formation by electrofusion of human B cells with heteromyeloma SPAM-8 cells. Hybridoma 14:265–269

Pavoni E, Monteriu G, Cianfriglia M, Minenkova O (2006) New display vector reduces biological bias for expression of antibodies in E. coli. Gene 391:120–129

Pereira-Smith OM, Robetorye S, Ning Y, Orson FM (1990) Hybrids from fusion of normal human T lymphocytes with immortal human cells exhibit limited life span. J Cell Physiol 144:546–549

Perkins S, Zimmermann U, Foung SK (1991) Parameters to enhance human hybridoma formation with hypoosmolar electrofusion. Hum Antibodies Hybridomas 2:155–159

Posner MR, Elboim H, Santos D (1987) The construction and use of a human-mouse myeloma analogue suitable for the routine production of hybridomas secreting human monoclonal antibodies. Hybridoma 6:611–625

Potera C (2005) Filling the niche for effective MAb screening. Genet Eng News 25

Pratt M, Mikhalev A, Glassy MC (1987) The generation of Ig-secreting UC 729-6 derived human hybridomas by electrofusion. Hybridoma 6:469–477

Ramirez JC, Tapia E, Esteban M (2002) Administration to mice of a monoclonal antibody that neutralizes the intracellular mature virus form of vaccinia virus limits virus replication efficiently under prophylactic and therapeutic conditions. J Gen Virol 83:1059–1067

Redfield RR, Wright DC, James WD, Jones TS, Brown C, Burke DS (1987) Disseminated vaccinia in a military recruit with human immunodeficiency virus (HIV) disease. N Engl J Med 316:673–676

Redmond MJ, Leyritz-Wills M, Winger L, Scraba DG (1986) The selection and characterization of human monoclonal antibodies to human cytomegalovirus. J Virol Methods 14:9–24

Rioux JD, Larose Y, Brodeur BR, Radzioch D, Newkirk MM (1994) Structural characteristics of four human hybridoma antibodies specific for the pp65 protein of the human cytomegalovirus and their relationship to human rheumatoid factors. Mol Immunol 31:585–597

Rodriguez WJ, Gruber WC, Welliver RC, Groothuis JR, Simoes EA, Meissner HC, Hemming VG, Hall CB, Lepow ML, Rosas AJ, Robertsen C, Kramer AA (1997) Respiratory syncytial virus (RSV) immune globulin intravenous therapy for RSV lower respiratory tract infection in infants and young children at high risk for severe RSV infections: Respiratory Syncytial Virus Immune Globulin Study Group. Pediatrics 99:454–461

Ruttmann E, Geltner C, Bucher B, Ulmer H, Hofer D, Hangler HB, Semsroth S, Margreiter R, Laufer G, Muller LC (2006) Combined CMV prophylaxis improves outcome and reduces the risk for bronchiolitis obliterans syndrome (BOS) after lung transplantation. Transplantation 81:1415–1420

Sa'adu A, Locniskar M, Bidwell D, Howard C, McAdam KP, Voller A (1992) Development and characterization of human anti-HBs antibodies. J Virol Methods 36:25–34

Saphire EO, Parren PW, Pantophlet R, Zwick MB, Morris GM, Rudd PM, Dwek RA, Stanfield RL, Burton DR, Wilson IA (2001) Crystal structure of a neutralizing human IGG against HIV-1: a template for vaccine design. Science 293:1155–1159

Sawyer LA (2000) Antibodies for the prevention and treatment of viral diseases. Antiviral Res 47:57–77

Schlom J, Wunderlich D, Teramoto YA (1980) Generation of human monoclonal antibodies reactive with human mammary carcinoma cells. Proc Natl Acad Sci U S A 77: 6841–6845

Schmidt E, Leinfelder U, Gessner P, Zillikens D, Brocker EB, Zimmermann U (2001) CD19+ B lymphocytes are the major source of human antibody-secreting hybridomas generated by electrofusion. J Immunol Methods 255:93–102

Sechet A, Bridoux F, Bauwens M, Ayache RA, Belmouaz S, Touchard G (2002) Prevention of cytomegalovirus infection and disease in high-risk renal transplant recipients with polyvalent intravenous immunoglobulins. Transplant Proc 34:812–813

Shay DK, Holman RC, Roosevelt GE, Clarke MJ, Anderson LJ (2001) Bronchiolitis-associated mortality and estimates of respiratory syncytial virus-associated deaths among US children, 1979–1997. J Infect Dis 183:16–22

Shirahata S, Katakura Y, Teruya K (1998) Cell hybridization, hybridomas, and human hybridomas. Methods Cell Biol 57:111–145

Shulman M, Wilde CD, Kohler G (1978) A better cell line for making hybridomas secreting specific antibodies. Nature 276:269–270

Siemoneit K, da Silva Cardoso M, Wolpl A, Koerner K, Subanek B (1994) Isolation and epitope characterization of human monoclonal antibodies to hepatitis C virus core antigen. Hybridoma 13:9–13

Small TN, Casson A, Malak SF, Boulad F, Kiehn TE, Stiles J, Ushay HM, Sepkowitz KA (2002) Respiratory syncytial virus infection following hematopoietic stem cell transplantation. Bone Marrow Transplant 29:321–327

Smith GL, Vanderplasschen A, Law M (2002) The formation and function of extracellular enveloped vaccinia virus. J Gen Virol 83:2915–2931

Smith LL, Coller HA, Roberts JM (2003) Telomerase modulates expression of growth-controlling genes and enhances cell proliferation. Nat Cell Biol 5:474–479

Snydman DR, Werner BG, Heinze-Lacey B, Berardi VP, Tilney NL, Kirkman RL, Milford EL, Cho SI, Bush HL Jr, Levey AS, et al (1987) Use of cytomegalovirus immune globulin to prevent cytomegalovirus disease in renal-transplant recipients. N Engl J Med 317:1049–1054

Sokos DR, Berger M, Lazarus HM (2002) Intravenous immunoglobulin: appropriate indications and uses in hematopoietic stem cell transplantation. Biol Blood Marrow Transplant 8:117–130

Stein LD, Sigal NH (1983) Limiting dilution analysis of Epstein-Barr virus-induced immunoglobulin production. Cell Immunol 79:309–319

Stewart SA, Hahn WC, O'Connor BF, Banner EN, Lundberg AS, Modha P, Mizuno H, Brooks MW, Fleming M, Zimonjic DB, Popescu NC, Weinberg RA (2002) Telomerase contributes to tumorigenesis by a telomere length-independent mechanism. Proc Natl Acad Sci USA 99:12606–12611

Stricker EA, Tiebout RF, Lelie PN, Zeijlemaker WP (1985) A human monoclonal IgG1 lambda anti-hepatitis B surface antibody. Production, properties, and applications. Scand J Immunol 22:337–343

Sugano T, Matsumoto Y, Miyamoto C, Masuho Y (1987) Hybridomas producing human monoclonal antibodies against varicella-zoster virus. Eur J Immunol 17:359–364

Sugano T, Tomiyama T, Matsumoto Y, Sasaki S, Kimura T, Forghani B, Masuho Y (1991) A human monoclonal antibody against varicella-zoster virus glycoprotein III. J Gen Virol 72:2065–2073

Tan MP, Koren G (2006) Chickenpox in pregnancy: revisited. Reprod Toxicol 21:410–420

Terrault NA, Vyas G (2003) Hepatitis B immune globulin preparations and use in liver transplantation. Clin Liver Dis 7:537–550

Terrault NA, Zhou S, McCory RW, Pruett TL, Lake JR, Roberts JP, Ascher NL, Wright TL (1998) Incidence and clinical consequences of surface and polymerase gene mutations in liver transplant recipients on hepatitis B immunoglobulin. Hepatology 28:555–561

Thompson JM, Lowe J, McDonald DF (1994) Human monoclonal anti-D secreting heterohybridomas from peripheral B lymphocytes expanded in the CD40 system. J Immunol Methods 175:137–140

Thompson WW, Shay DK, Weintraub E, Brammer L, Cox N, Anderson LJ, Fukuda K (2003) Mortality associated with influenza and respiratory syncytial virus in the United States. JAMA 289:179–186

Tian C, Luskin GK, Dischert KM, Higginbotham JN, Shepherd BE, Crowe JE Jr (2007) Evidence for preferential Ig gene usage and differential TdT and exonuclease activities in human naive and memory B cells. Mol Immunol 44:2173–2183

Tiebout RF, van Boxtel-Oosterhof F, Stricker EA, Zeijlemaker WP (1987) A human hybrid hybridoma. J Immunol 139:3402–3405

Traggiai E, Becker S, Subbarao K, Kolesnikova L, Uematsu Y, Gismondo MR, Murphy BR, Rappuoli R, Lanzavecchia A (2004) An efficient method to make human monoclonal antibodies from memory B cells: potent neutralization of SARS coronavirus. Nat Med 10:871–875

Ueki Y, Goldfarb IS, Harindranath N, Gore M, Koprowski H, Notkins AL, Casali P (1990) Clonal analysis of a human antibody response. Quantitation of precursors of antibody-producing cells and generation and characterization of monoclonal IgM, IgG, and IgA to rabies virus. J Exp Med 171:19–34

Vaisbourd M, Ignatovich O, Dremucheva A, Karpas A, Winter G (2001) Molecular characterization of human monoclonal antibodies derived from fusions of tonsil lymphocytes with a human myeloma cell line. Hybrid Hybridomics 20:287–292

Varga M, Remport A, Hidvegi M, Peter A, Kobori L, Telkes G, Fazakas J, Gerlei Z, Sarvary E, Sulyok B, Jaray J (2005) Comparing cytomegalovirus prophylaxis in renal transplantation: single center experience. Transpl Infect Dis 7:63–67

Welliver RC (2003) Respiratory syncytial virus and other respiratory viruses. Pediatr Infect Dis J 22: S6–S10; discussion S10–S12

Winter G, Griffiths AD, Hawkins RE, Hoogenboom HR (1994) Making antibodies by phage display technology. Annu Rev Immunol 12:433–455

Xiao XM, Li AZ, Chen X, Zhu YK, Miao J (2007) Prevention of vertical hepatitis B transmission by hepatitis B immunoglobulin in the third trimester of pregnancy. Int J Gynaecol Obstet 96:167–170

Xu JL, Davis MM (2000) Diversity in the CDR3 region of V(H) is sufficient for most antibody specificities. Immunity 13:37–45

Xu Q, Xiao L, Lu XB, Zhang YX, Cai X (2006) A randomized controlled clinical trial: interruption of intrauterine transmission of hepatitis B virus infection with HBIG. World J Gastroenterol 12:3434–3437

Yoshinari K, Arai K, Kimura H, Matsumoto K, Yamaguchi Y (1996) Efficient production of IgG human monoclonal antibodies by lymphocytes stimulated by lipopolysaccharide, pokeweed mitogen, and interleukin 4. In Vitro Cell Dev Biol Anim 32:372–377

Young J (2002) Development of a potent respiratory syncytial virus-specific monoclonal antibody for the prevention of serious lower respiratory tract disease in infants. Respir Med 96 [Suppl B]: S31–35

Zhu Y, Jin B, Sun C, Huang C, Liu X (1993) The effects of hybridoma growth factor in conditioned media upon the growth, cloning, and antibody production of heterohybridoma cell lines. Hum Antibodies Hybridomas 4:31–35

Zhu Y, Jin B, Jiang S, Sun K, Sun C, Liu X (1999) Improved fusion partners transfected with DNA fragment encoding IL-11 on generation of human B lymphocyte hybridomas. Hum Antibodies 9:1–7

Zikos P, Van Lint MT, Lamparelli T, Gualandi F, Occhini D, Mordini N, Berisso G, Bregante S, Bacigalupo A (1998) A randomized trial of high dose polyvalent intravenous immunoglobulin (HDIgG) vs. Cytomegalovirus (CMV) hyperimmune IgG in allogeneic hemopoietic stem cell transplants (HSCT). Haematologica 83:132–137

Index

Current Topics in Microbiology and Immunology

Volumes published since 1989

Vol. 295: **Sullivan, David J.; Krishna Sanjeew (Eds.):** Malaria: Drugs, Disease and Post-genomic Biology. 2005. 40 figs., XI, 446 pp. ISBN 3-540-25363-7

Vol. 296: **Oldstone, Michael B. A. (Ed.):** Molecular Mimicry: Infection Induced Autoimmune Disease. 2005. 28 figs., VIII, 167 pp. ISBN 3-540-25597-4

Vol. 297: **Langhorne, Jean (Ed.):** Immunology and Immunopathogenesis of Malaria. 2005. 8 figs., XII, 236 pp. ISBN 3-540-25718-7

Vol. 298: **Vivier, Eric; Colonna, Marco (Eds.):** Immunobiology of Natural Killer Cell Receptors. 2005. 27 figs., VIII, 286 pp. ISBN 3-540-26083-8

Vol. 299: **Domingo, Esteban (Ed.):** Quasispecies: Concept and Implications. 2006. 44 figs., XII, 401 pp. ISBN 3-540-26395-0

Vol. 300: **Wiertz, Emmanuel J.H.J.; Kikkert, Marjolein (Eds.):** Dislocation and Degradation of Proteins from the Endoplasmic Reticulum. 2006. 19 figs., VIII, 168 pp. ISBN 3-540-28006-5

Vol. 301: **Doerfler, Walter; Böhm, Petra (Eds.):** DNA Methylation: Basic Mechanisms. 2006. 24 figs., VIII, 324 pp. ISBN 3-540-29114-8

Vol. 302: **Robert N. Eisenman (Ed.):** The Myc/Max/Mad Transcription Factor Network. 2006. 28 figs., XII, 278 pp. ISBN 3-540-23968-5

Vol. 303: **Thomas E. Lane (Ed.):** Chemokines and Viral Infection. 2006. 14 figs. XII, 154 pp. ISBN 3-540-29207-1

Vol. 304: **Stanley A. Plotkin (Ed.):** Mass Vaccination: Global Aspects – Progress and Obstacles. 2006. 40 figs. X, 270 pp. ISBN 3-540-29382-5

Vol. 305: **Radbruch, Andreas; Lipsky, Peter E. (Eds.):** Current Concepts in Autoimmunity. 2006. 29 figs. IIX, 276 pp. ISBN 3-540-29713-8

Vol. 306: **William M. Shafer (Ed.):** Antimicrobial Peptides and Human Disease. 2006. 12 figs. XII, 262 pp. ISBN 3-540-29915-7

Vol. 307: **John L. Casey (Ed.):** Hepatitis Delta Virus. 2006. 22 figs. XII, 228 pp. ISBN 3-540-29801-0

Vol. 308: **Honjo, Tasuku; Melchers, Fritz (Eds.):** Gut-Associated Lymphoid Tissues. 2006. 24 figs. XII, 204 pp. ISBN 3-540-30656-0

Vol. 309: **Polly Roy (Ed.):** Reoviruses: Entry, Assembly and Morphogenesis. 2006. 43 figs. XX, 261 pp. ISBN 3-540-30772-9

Vol. 310: **Doerfler, Walter; Böhm, Petra (Eds.):** DNA Methylation: Development, Genetic Disease and Cancer. 2006. 25 figs. X, 284 pp. ISBN 3-540-31180-7

Vol. 311: **Pulendran, Bali; Ahmed, Rafi (Eds.):** From Innate Immunity to Immunological Memory. 2006. 13 figs. X, 177 pp. ISBN 3-540-32635-9

Vol. 312: **Boshoff, Chris; Weiss, Robin A. (Eds.):** Kaposi Sarcoma Herpesvirus: New Perspectives. 2006. 29 figs. XVI, 330 pp. ISBN 3-540-34343-1

Vol. 313: **Pandolfi, Pier P.; Vogt, Peter K. (Eds.):** Acute Promyelocytic Leukemia. 2007. 16 figs. VIII, 273 pp. ISBN 3-540-34592-2

Vol. 314: **Moody, Branch D. (Ed.):** T Cell Activation by CD1 and Lipid Antigens, 2007, 25 figs. VIII, 348 pp. ISBN 978-3-540-69510-3

Vol. 315: **Childs, James, E.; Mackenzie, John S.; Richt, Jürgen A. (Eds.):** Wildlife and Emerging Zoonotic Diseases: The Biology, Circumstances and Consequences of Cross-Species Transmission. 2007. 49 figs. VII, 524 pp. ISBN 978-3-540-70961-9

Vol. 316: **Pitha, Paula M. (Ed.):** Interferon: The 50th Anniversary. 2007. VII, 391 pp. ISBN 978-3-540-71328-9

Printing: Krips bv, Meppel
Binding: Stürtz, Würzburg